A-LEVEL GEOGRAPHY
TOPIC MASTER

CHANGING PLACES

Simon Oakes

Series editor
Simon Oakes

HODDER
EDUCATION
AN HACHETTE UK COMPANY

To Hal, Ned and Rosalind

The Publishers would like to thank the following for permission to reproduce copyright material.

Acknowledgements

Photo credits can be found on page 232

Thanks to the Geographical Association for permission to reproduce content from Chapters 3 and 4 of Rawlings Smith, E,. Oakes, S. and Owens, A. (2016) Top Spec Geography: Changing Places. Sheffield: Geographical Association.

p. 16 Figure 1.8 Lorenz curves showing the level of segregation for different ethnic communities in the UK in the 1991 Census. Republished with permission of JOHN WILEY & SONS, from [Transactions of the Institute of British Geographers, New Series, Vol. 21, No. 1 (1996), pp. 216-235]; permission conveyed through Copyright Clearance Center, Inc.; **p. 17 Figure 1.9** Graphic by Billy Ehrenberg - Shannon and Graham Parrish from *Northern Powerhouse project threatened by 'brain drain'* by Andrew Bounds and Chris Tighe. Financial Times 18th April 2016; **p. 22 Figure 1.13** Donald G. Janelle (2010), CENTRAL PLACE DEVELOPMENT IN A TIME-SPACE FRAMEWORK, The Professional Geographer, 20 (1): 5-10 © The Association of American Geographers, www.aag.org, reprinted by permission of Taylor & Francis Ltd, httpwww.tandfonline.com on behalf of The Association of American Geographers. **p. 24 Figure 1.14** From *How laundered money shapes London's property market* by Judith Evans. Financial Times 6th April 2016; **p.30** Construction in turning London into a city of holes by Edwin Heathcote. Financial Times Magazine 23rd April 2016; **p.70 Figure 2.26** Reprinted with permission of The Institute of Cultural Capital, University of Liverpool; **p. 102 Figure 3.20** Graphic by John Burn-Murdoch from *Left behind: can anyone save the towns the economy forgot* by Sarah O'Connor. Financial Times Magazine. November 18th 2017; **p.113 Table 3.9** Reprinted with permission of City AM Ltd; **p.120 Figure 4.3** Reproduced with permission from Professor Alastair Owens; **p. 130 Figure 4.12** Graphic by John Burn-Murdoch from *Young professionals resist London's lure and head north for jobs* by Andrew Bounds. Financial Times 11th May 2017; **p. 172 Figure 5.19** Graphic by Alan Smith from *Commuting times and housing costs compared in eight major cities* by Hugo Cox and Alan Smith. Financial Times. 18th February 2016

Index supplied by LNS Indexing

Every effort has been made to trace all copyright holders, but if any have been inadvertently overlooked, the Publishers will be pleased to make the necessary arrangements at the first opportunity.

Although every effort has been made to ensure that website addresses are correct at time of going to press, Hodder Education cannot be held responsible for the content of any website mentioned in this book. It is sometimes possible to find a relocated web page by typing in the address of the home page for a website in the URL window of your browser.

Hachette UK's policy is to use papers that are natural, renewable and recyclable products and made from wood grown in well-managed forests and other controlled sources. The logging and manufacturing processes are expected to conform to the environmental regulations of the country of origin.

Orders: please contact Hachette UK Distribution, Hely Hutchinson Centre, Milton Road, Didcot, Oxfordshire, OX11 7HH. Telephone: +44 (0)1235 827827. Email education@hachette.co.uk Lines are open from 9 a.m. to 5 p.m., Monday to Friday. You can also order through our website: www.hoddereducation.co.uk

ISBN: 978 1 5104 2753 2

© Simon Oakes 2018

First published in 2018 by

Hodder Education,

An Hachette UK Company

Carmelite House

50 Victoria Embankment

London EC4Y 0DZ

www.hoddereducation.com

Impression number   10 9 8 7 6 5 4 3 2

Year   2022

All rights reserved. Apart from any use permitted under UK copyright law, no part of this publication may be reproduced or transmitted in any form or by any means, electronic or mechanical, including photocopying and recording, or held within any information storage and retrieval system, without permission in writing from the publisher or under licence from the Copyright Licensing Agency Limited. Further details of such licences (for reprographic reproduction) may be obtained from the Copyright Licensing Agency Limited, www.cla.co.uk

Cover photo © Alija / E+ / Getty Images

Illustrations by Barking Dog Art

Typeset in India by Aptara Inc.

Printed and bound by CPI Group (UK) Ltd, Croydon, CR0 4YY

A catalogue record for this title is available from the British Library.

Contents

Introduction 4

CHAPTER 1: Place characteristics, dynamics and connections 5

1 Place characteristics 5
2 Place dynamics 19
3 Place networks and layered connections 26
4 Evaluating the issue: to what extent can places be completely protected from change? 32

CHAPTER 2: Place meanings, representations and experiences 39

1 Place meanings for individuals and societies 39
2 The power of place representations 48
3 Representations of the city and the countryside in popular culture 58
4 Evaluating the issue: to what extent can place meanings and representations become a cause of conflict? 66

CHAPTER 3: Place changes, challenges and inequalities 74

1 Deindustrialisation and the cycle of deprivation 74
2 Twenty-first-century economic, political and technological challenges 86
3 Changing demographic and cultural characteristics of places 98
4 Evaluating the issue: assessing the severity of spatial inequalities in the UK 109

CHAPTER 4: The place-remaking process 117

1 Place-remaking approaches, strategies and players 117
2 Cultural heritage and place-remaking processes 131
3 Remaking contemporary places 136

4 Evaluating the issue: assessing the importance of different players in the place-remaking process 143

CHAPTER 5: Creating sustainable places 152

1 Government policies for economically sustainable places 152
2 Encounters with cultural and social differences and inequalities 161
3 Tackling urban environmental stress 170
4 Evaluating the issue: to what extent have actions to create sustainable places been successful? 175

CHAPTER 6: Issues for rural places 184

1 Rural places, players and connections 184
2 Change and challenge in the differentiated countryside 190
3 Place remaking in a rural context 195
4 Evaluating the issue: Discussing differing views about the identity of rural places 200

CHAPTER 7: Study guides 208

1 AQA A-level Geography: Changing Places and Contemporary Urban Environments 208
2 Pearson Edexcel A-level Geography: Shaping Places (either Regenerating Places or Diverse Places) 213
3 OCR A-level Geography: Changing Spaces; Making Places 218
4 WJEC and Eduqas A-level Geography: Changing Places 222

Index 227

Human geography at A-level has changed radically in recent years. These alterations in part reflect how university geography courses are evolving over time. New topics first introduced to the undergraduate curriculum in the 1990s have since become mainstream degree-level topics. Foremost among these is the study of 'changing places'. In 2016, at the insistence of an expert advisory panel, it became a required topic for A-level Geography specifications too. This book aims to foster A-level student learning about changing places using a balance of theory (conceptual thinking about place) and thought-provoking case studies. The latter are mainly focused on places in the UK but on occasion look further afield (for example to Barcelona, which features on the cover of this book).

The A-level Geography Topic Master series

The books in this series are designed to support learners who aspire to reach the highest grades. To do so requires more than learning-by-rote. Only around one-third of available marks in an A-level Geography examination are allocated to the recall of knowledge (*assessment objective 1, or AO1*). A greater proportion is reserved for higher-order cognitive tasks, including the **analysis, interpretation** and **evaluation** of geographic ideas and information (*assessment objective 2, or AO2*). Therefore, the material in this book has been purposely written and presented in ways which encourage active reading, reflection and critical thinking. The overarching aim is to help you develop the analytical and evaluative 'geo-capabilities' needed for examination success. Opportunities to practise and develop **data manipulation skills** are also embedded throughout the text (supporting *assessment objective 3, or AO3*).

All *Geography Topic Master* books prompt students constantly to 'think geographically'. In practice this can mean learning how to seamlessly integrate **geographic concepts** – including place, scale, interdependency, causality and inequality – into the way we think, argue and write. The books also take every opportunity to establish **synoptic links** (this means making 'bridging' connections between themes and topics). Frequent page-referencing is used to create links between different chapters and sub-topics. Additionally, numerous connections have been highlighted between *Changing places* and other separate geography topics, such as *Global systems* or *Carbon and water cycles*.

Using this book

The book may be read from cover to cover since there is a logical progression between chapters (each of which is divided into four sections). On the other hand, a chapter may be read independently whenever required as part of your school's scheme of work for this topic. A common set of features are used in each chapter:

- *Aims* establish the four main points (and sections) of each chapter
- *Key concepts* are important ideas relating either to the discipline of Geography as a whole or more specifically to the study of *Changing places*
- *Contemporary case studies* apply geographical ideas, theories and concepts to real-world local contexts that have often been impacted upon by external global forces (such as the global financial crisis, China's growth and Brexit)
- *Analysis and interpretation* features help you develop the geographic skills and capabilities needed for the application of knowledge and understanding (AO2), data manipulation (AO3) and, ultimately, exam success
- *Evaluating the issue* brings each chapter to a close by discussing a key *Changing places* issue (typically involving competing perspectives and views)
- Also included at the end of each chapter are the *Chapter summary, Refresher questions, Discussion activities, Fieldwork focus* (supporting the independent investigation) and selected *Further reading*.

Place characteristics, dynamics and connections

The dynamic relationship between people, economies and the physical environment helps create multi-layered places. Different places develop distinctive identities over time. Using a range of ideas, concepts and supporting data, this chapter:

- analyses the different human and physical elements which interconnect to give a place character
- investigates how places change over time on account of dynamic internal and external processes
- explores ways in which places and communities are shaped by past and present network connections and relationships with other places at regional, national and global scales
- evaluates the extent to which places can ever be preserved and protected from agents of change such as globalisation.

KEY CONCEPTS

Place identity The physical and human elements that help to make a place distinctive from other places. This chapter examines the quantifiable physical, economic and demographic characteristics of places (whereas Chapter 2 explores more subjective interpretations of place identity).

Interdependence The relations of mutual dependence that develop between different places over time. Also, the influence that the society, economy and landscape of a place exert over one another.

Globalisation The intensification of connections between different places on a global scale. Accelerating flows of capital, commodities, people and information are the result of a 'shrinking world' shaped by markets, technology and political changes. Some people and societies embrace globalisation; others try to reject it.

 KEY TERMS

Space The basic organising concept of the geographer – geographical studies historically aimed to establish the locations of people and phenomena on the Earth's surface and represent this knowledge to others using maps. In contemporary studies of places, space can be understood as the distance which separates places. Two important points arise. First, the space which separates two places is never empty or devoid of features and meaning. For example, migrants will sometimes pass through a place they did not know about previously while moving from their home place towards an intended destination. They may decide to settle at this 'intervening opportunity' instead of completing their journey. Second, the space (or distance) between places is both real and perceived: the sense of distance between places lessens wherever fast transport and communications are available.

 Place characteristics

▶ *What are the main elements which make up a place, and how do they combine to create a distinctive identity?*

Discovering places

A place is a portion of geographic **space** the identity of which is viewed as being distinctive in some way. Particular places have unique landscapes deriving from the underlying physical geography as well as the way different societies have shaped their surface appearance over time. 'Layered history' is an important aspect of a place's character. This term describes the

▲ **Figure 1.1** Clerkenwell in London: this central area of the city is characterised by an urban mosaic of contrasting buildings and land uses from different historical eras

 KEY TERMS

Near places In real terms, these are places located in adjacent settlements and the wider surrounding region. In the UK, historical rural–urban migration flows took place between cities and their surrounding countryside. The term can also be used to describe places that *feel* close-at-hand thanks to technology and transport – even though the real-world distance is great.

Far places Distant places within a country, or places in other countries often at a considerable distance away. Also, isolated places that feel distant because they take a long time to reach (even though the real-world distance may be small).

Region A broad area, such as the Midlands or Lake District, within which places share certain physical and cultural characteristics.

accumulated landscape imprint of physical and human processes operating in the past and present. Some of the UK's oldest cities, such as Chester, London, Bristol and York, are built on a floodplain or coastline. In each case, evidence of past flooding or erosion is demonstrated by the defences that have become an important part of their modern place identity. Older places found within these cities, such as London's Clerkenwell district, are characterised by an 'urban mosaic' of contrasting buildings and land uses (see Figure 1.1). These include Victorian and contemporary housing, Norman churches, medieval synagogues, postwar mosques and repurposed factories now functioning as offices, nightclubs or restaurants. This montage of ancient and modern land uses is the outcome of a long, multi-layered history of economic change and challenge, population growth and migration.

The present-day landscape of Clerkenwell offers many clues about how this place's identity – or 'personality' – has been reshaped repeatedly by its changing relationships with other places and societies at a range of geographic scales. All places are dynamic and 'relational' to some extent. This is because the society and economy which have made them, and upon which they depend, are themselves in a constant state of flux. In part, this dynamism is a result of changing connections with other **near places** (at the local and national scale) and **far places** (at the international and global scale). In recent decades, globalisation has accelerated the rate of change experienced by many places, including Clerkenwell, where an increasing number of properties, businesses and shops are now foreign-owned. Political forces are another highly significant influence on the extent to which places change (or do not) over time (see pages 32–35).

Place and scale

Villages, small towns and city neighbourhoods are all local places. Each, in turn, is embedded in a larger-scale geographical context, such as a **region** or city. The inner-city district of Bootle and the fringe village of Formby both belong to the city of Liverpool in the northwest of England, for instance. All these entities – Bootle, Formby, Liverpool and the northwest – can be understood as places, insofar as each possesses a set of physical and human features that provide an identity and distinguish them from other places at the same scale.

The idea of large-scale 'regional identity' was of central importance for early twentieth-century geographical thought. Writing in the early 1900s, Paul Vidal de la Blache introduced the concept of homogeneous regions – that is, broad areas within which a particular 'bond' had developed between the natural landscape and the societies living there. Vidal's ideas survive in the way we view the different regions or countries as possessing a combination of distinctive physical and cultural characteristics. The characteristic mills, factory chimneys and farm walls of the Yorkshire region were built with local Pennine gritstone, for instance (see Figure 1.2). They give Yorkshire a sense of uniformity, along with its regional dialect. Yet each village or urban neighbourhood within Yorkshire is also a distinctive place in its own right at the local scale.

Two important principles follow from this. First, part of a particular place's identity may be shared by the wider region or city to which that place belongs. Second, the economic decline of particular places is often part of a much bigger picture of sectoral and regional decline (see Chapters 3 and 4).

At the very largest geographic scale of global systems or world development studies, the word 'place' is used sometimes as a synonym for 'country' by non-geographers and geographers alike. For instance, an exam question might ask a student to discuss why levels of economic development vary 'from place to place' across the world. In this context it is reasonable to give credit to the student who uses two countries at different stages of development as examples of 'place'. A good answer may do this before 'drilling down' to a far more local level and comparing the economic development of two neighbouring villages within a region.

To conclude, the word 'place' can be used legitimately in several different ways, according to the geographic scale of enquiry that is being undertaken. But for practical (fieldwork) reasons, a place is best understood in A-level Geography as being *a distinctive locality at a geographical scale somewhere between a street and a city or region.* Accordingly, the word 'place' is used primarily in that sense in this book. The places written about here are mostly villages, small towns and local city neighbourhoods. Although Chapters 4 and 5 deal with the redevelopment of cities such as Liverpool, Manchester, Birmingham and London, the focus is invariably *city centre* redevelopment. These city centres can themselves be viewed as local places embedded within much larger-scale settlements.

Urban and rural places

This book is concerned equally with the study of urban and rural places, and the changes and challenges they experience. Both urban and rural studies are significant sub-disciplines within Geography. Each has its own highly specialised vocabulary. Some of this terminology needs clarifying, including the terms 'rural' and 'urban' themselves, both of which can be defined in varying ways. The rural–urban distinction may, for instance, be based primarily on 'form', that is, what an area looks like in terms of its landscape features and housing density. Alternatively, a classification can be made based on the economic function of the land, namely the services and employment found there. Function-based definitions have changed over time, however, because it no longer makes sense to define a rural area as being one where agricultural employment dominates. Although this may once have been the case, today very few rural people depend on farming to make a living. Rural functions have diversified in recent

▲ **Figure 1.2** Hebden Bridge has its own distinctive identity while sharing certain landscape characteristics with much of Yorkshire as a whole

 KEY TERMS

Urban In its most general sense, this word relates to towns, cities and the life that is lived there. In some countries and contexts, it may have a more specific meaning (usually in relation to land use or population).

Rural In its most general sense, this word relates to the countryside and the life that is lived there. In some countries and contexts it may have a more specific meaning (usually in relation to land-use or population).

Function The role(s) a place or settlement plays in its local community as well as the wider world. The main function of a small town may be to provide supermarket services, for instance. Large cities that are home to universities and the headquarters of large companies may offer national or even global functions.

years to embrace tourism, technology and leisure services (see pages 139 and 195–199).

In the UK today, it is settlement population size – rather than either form or function – which policymakers use as the main marker of difference between rural and urban areas. In total, six rural and four urban 'types' are identified by the UK Government (see Figure 1.3). The most important guidelines underpinning this classification are that:

- 'urban' refers to an individual settlement of more than 10,000 people
- 'rural' refers to open countryside or areas where people live in smaller settlements of fewer than 10,000 people (small market towns, villages and hamlets).

In practice, using these terms is not always straightforward in the study of Geography. First, people living in small towns are sometimes defined as a rural population, which may seem counterintuitive. Second, some British

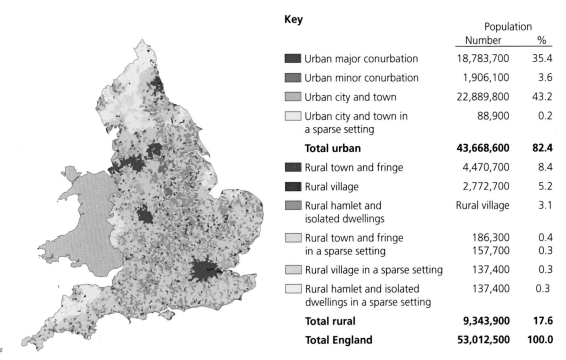

Key	Population	
---	Number	%
Urban major conurbation	18,783,700	35.4
Urban minor conurbation	1,906,100	3.6
Urban city and town	22,889,800	43.2
Urban city and town in a sparse setting	88,900	0.2
Total urban	**43,668,600**	**82.4**
Rural town and fringe	4,470,700	8.4
Rural village	2,772,700	5.2
Rural hamlet and isolated dwellings	Rural village	3.1
Rural town and fringe in a sparse setting	186,300 157,700	0.4 0.3
Rural village in a sparse setting	137,400	0.3
Rural hamlet and isolated dwellings in a sparse setting	137,400	0.3
Total rural	**9,343,900**	**17.6**
Total England	**53,012,500**	**100.0**

▲ **Figure 1.3** Rural and urban areas of England according to the official ten-category classification system. Source: 2011 UK Census

counties and regions are characterised as 'rural regions' within a national context. Examples include Cornwall or the Scottish Highlands. Yet these rural regions include large urban settlements of more than 10,000 people, such as St Ives (Cornwall) and Inverness (the Highlands). Similarly, urban regions such as Merseyside and Greater Manchester include small rural settlements located in the green belt surrounding their major cities.

Finally, it would be wrong to infer from Figure 1.3 that rural and urban places always have boundaries that can be clearly identified using field evidence. In fact, visible landscape characteristics may not necessarily correlate at all well with the 'official' picture provided by administrative boundary maps. Urban areas rarely terminate neatly at a city wall beyond which green fields stretch to the horizon. Instead, there is often a gradual change in housing density due to urban sprawl at the rural–urban fringe. Housing slowly thins out as gardens get bigger; residential estates start to become interspersed with golf courses, reservoirs, landfill sites or small areas of woodland. Ribbon development along transport arteries results in areas which are predominantly countryside yet have veins of housing or industry running through them.

In fringe areas, urban and rural forms and functions become fused together. These are hybrid places where farm workers and city commuters are sometimes neighbours. A range of specialist vocabulary exists for these 'liminal' areas which lack a clearly urban or rural character. Words and phrases like 'rurban' places, the rural–urban continuum and edgelands are all used in the study of overlapping rural and urban populations and land uses at the edges of major settlements (see Chapter 6).

Home places

This chapter explores the characteristics, dynamics and connections of the places which geographers can discover, map and analyse. One way to begin is through the study of your home neighbourhood or where you study. This is your home place. Think about how the place where you live has changed during your own lifetime or that of older residents who you know, including family members. Even recently-built housing estates may have changed in significant ways during their relatively short lives due to migration and economic restructuring. UK new towns dating from the 1950s and 1960s – including Milton Keynes (see Figure 1.4) and Stevenage – also contain traces of much earlier settlement if you look carefully.

🔑 **KEY TERMS**

Green belt A land use designation which forms an important part of planning law. The green belt is a 'girdle' of undeveloped land which encircles towns and cities in the UK (although it contains some villages and other developments built prior to the introduction of the green belt).

Urban sprawl The outward expansion of a settlement, as people and economic activities relocate near its edges.

Rural–urban fringe A zone of change between the continuously built-up suburbs and the surrounding countryside.

Rural–urban continuum The unbroken transition from sparsely populated or unpopulated rural places to densely populated and intensively used urban places.

Edgelands Transitional areas of land found where countryside borders the town or city. Lacking either a truly rural or urban character, these are hybrid (or 'liminal') places possessing their own unique identity.

Home place Your home neighbourhood or the place where you study.

▲ **Figure 1.4** Milton Keynes is a purpose-built settlement dating from the 1960s which has nonetheless experienced major changes in its short lifetime

KEY TERM

Cultural landscape The distinctive character of a geographical place or region which has been shaped over time by a combination of physical and human processes. In Geography, the concept is associated strongly with the work of Carl Sauer.

What's in a place?

Figure 1.5 provides one possible framework for the study of place characteristics. This approach draws on several well-established geographical concepts: (i) physical site factors, (ii) economic functions and (iii) the cultural landscape (which includes a place's population characteristics). Figure 1.5 also emphasises how part of a place's identity is gained from the connections and relations it has with other places.

The site of a place

The site of a place is the actual land it is built on. Settlements have historically taken root wherever geographical site factors favour economic activities that cannot be carried out as profitably in other locations. Expressed simply, local resources such as coal or water explain why some places are where they are. Physical geography helps to shape the characteristics of entire cities and regions. For example, in Sheffield and the surrounding region of Hallamshire, the iron and steel industries traditionally created a strong sense of place, whereas coal moulded the identity of settlements in South Wales.

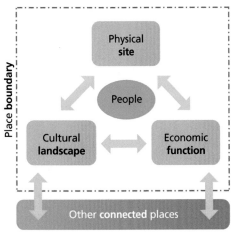

▲ **Figure 1.5** The main elements of place

Particular places and neighbourhoods within large cities have their own advantageous site factors and topography. The high elevation of London's Hampstead and Highgate neighbourhoods provided wealthy Victorians with clean air and safe water. They remain affluent places today (see page 18).

In addition to the site itself, climate is another important physical influence on regions and places. It determines what plants and crops can grow. France's Champagne region enjoys warm, dry summers. Grapes used for viticulture (wine-making) thrive in the flinty, chalk soils found there. The world-famous flavour of its wines is said to reflect these geographic characters of Champagne's climate and geology in equal measure. Physical geography provides essential elements of place identity for villages in the Champagne region, such as Chouilly and Cramant.

However, it is important to not overstate the extent to which physical geography decides the ways in which a place (and society) can or cannot develop over time. The prevailing view within human geography is that human ingenuity, technology and political forces usually have the final say about what happens in any one place. Environmental 'obstacles' (such as lack of water) can often be overcome if people have the will (and money) to make it happen.

The economic functions of places

The economic function of a place is what it does in terms of providing services and work for people. Originally, this was often linked closely with site factors. Table 1.1 shows the traditional economy of three large UK cities.

Pearson Edexcel

AQA

OCR

WJEC/Eduqas

City	Traditional economy (1700–1950)	Urban cultural landscape features
London	London is seven times larger (in population size) than any other UK city and has an accordingly diverse economy. Traditionally, docks, textile and furniture making, food and drink processing, munitions and engineering were important employers.	A wide range of local cultures, from the Cockney East End to Chelsea and Bloomsbury.
Birmingham	In the twentieth century, Birmingham became a hub for the British car industry (important employers included Austin and Dunlop). Earlier success came from jewellery, guns and food processing (Cadbury's is a Birmingham brand).	A rich music heritage, ranging from the work of Elgar to the city's 'heavy metal' scene.
Glasgow	In riverside places such as Dumbarton, Clydebank and Govan Graving Docks, shipbuilding was a major source of wealth for Glasgow. So too were fishing, textiles and manufacturing, sugar and tobacco processing and engineering.	The Glaswegian dialect; traditional religious rivalry between Celtic and Rangers football clubs.

▲ **Table 1.1** The traditional economy of UK cities and the linked culture of urban places

As we have already seen, most cities, towns and villages are located where they are for economic reasons. The comparative advantages enjoyed by different settlements can be linked to site factors such as soils, water supply or proximity to raw materials. Many settlements have, however, changed their function over time. Liverpool and Manchester are now **post-industrial** cities where consumer services have replaced manufacturing industries. As a result, many places within these cities have changed beyond all recognition in recent decades (see Chapter 4). In some **post-productive** rural places, agriculture has given way to tourism.

The cultural landscape of urban places

The cultural landscape is everything we experience in a place. It is the totality of the changes which people have brought to the natural landscape, including the architecture, infrastructure and demography of a place. It also includes the art, music (part of a place's **soundscape**) and sporting life of somewhere. Alongside their traditional industries, Table 1.1 also shows selected cultural landscape traits which have developed in those cities. Before car ownership and mass public transport, many urban workers lived their lives in close-knit local neighbourhoods and knew of little beyond this. Low pay, long working weeks and short holidays meant few people travelled to other places. Leisure time was in limited supply and involved pursuing activities near the workplace. As a result, urban cultural landscapes developed which strongly reflected each settlement's economic functions. City football teams originally drew their amateur players from

  KEY TERMS

Post-industrial An economy or society where traditional manufacturing or mining employment has been replaced by an employment structure focused on services and technology. A post-industrial city is a settlement where most jobs are in the tertiary and quaternary sectors.

Post-productive A rural place or economy where agriculture is no longer a major employer (although large areas of land may still be used for mechanised agriculture).

Soundscape The natural and human-made sounds that are generated by a particular environment and help shape people's experience of a place.

DERBY COUNTY

▲ **Figure 1.6** Football club badges continue to show links with the past economies of places. Can you deduce what the traditional economic functions were for each of these cities or regions?

local factories. For instance, the cannon on Arsenal's badge reflects the club's birth in the 1880s among the munitions factories of Woolwich, sited by the River Thames. By 1900, London supported over 100 local football teams, each one rooted in a different factory neighbourhood (see Figure 1.6). In contrast, Welsh coal and slate mining communities often became renowned for their male voice choirs.

North American cities also have rich cultural histories linked to their economic traditions. For example, in the 1950s, Detroit was a national hub for car making where major transnational corporations (TNCs), including Ford, were based. After the successes of the civil rights movement in the 1960s, large numbers of African-Americans had migrated there from southern states in search of manufacturing work. Detroit subsequently became the home of Tamla Motown, the record label that launched Michael Jackson, Stevie Wonder and many more black music acts in the 1960s and 1970s. Motown is an abbreviation of 'Motor Town', a clear illustration of how the economy of Detroit shaped its cultural landscape, particularly places like the Grand Boulevard district where Motown was first headquartered. Equally important was the role played by Detroit's demography and in-migration from other places: Motown artists drew on black musical traditions from the southern states, including the Blues, a musical form whose roots span the Atlantic to West Africa.

Not all people view the cultural landscape of the urban place where they live positively, however. Between 2003 and 2013, a series of best-selling humorous books titled *Crap Towns* documented 'the worst places to live in the UK' as voted for by members of the public visiting *The Idler* website. Popular as Christmas stocking fillers, these volumes were controversial because of their perceived harmful effects on real patterns of investment in the places they mocked. Table 1.2 shows the top ten rankings from the very first *Crap Towns* volume in 2003. Local newspapers and politicians were quick to defend these locations; later in this book, you will encounter regeneration and redevelopment work carried out subsequently in Hull, Liverpool and Hackney.

Rank	Location
1	Hull
2	Cumbernauld
3	Morecambe
4	Hythe
5	Winchester
6	Liverpool
7	St Andrews
8	Bexhill-on-Sea
9	Basingstoke
10	Hackney, London

▲ **Table 1.2** The 2003 top ten *Crap Towns* rankings

The cultural landscape of rural places

Many rural places have distinctive characteristics which derived originally from each local community's agricultural practices or from artisan and craft traditions which had often developed alongside farming. Making cheese or knitting woollen garments was a good way for agricultural communities to add value to their produce, for instance. In other places, local ores were used for metalworking by blacksmiths. Prior to large-scale manufacturing in the 1700s, rural crafts were vital to the UK's economy during the 'proto-industrial' development phase of the 1600s. Because many villages were relatively isolated prior to modern transport, unique craft-making and cultural traditions sometimes developed in such places. These ways of life did not always survive the Industrial Revolution, however. Out-migration meant traditions were no longer handed down from generation to generation.

Where they have survived, rural traditions sometimes include unique words and language. The geographer Robert Macfarlane (2015) has uncovered hundreds of different local dialect words for 'rain' or 'water' while carrying out his extensive studies of language in rural places that used to be isolated. Local communities often had unique music too, which researchers have worked hard to preserve.

- In the 1800s, songs written by people in isolated communities were often passed from generation to generation *in situ* but remained unknown to people living outside of these places. The melody for 'O Little Town of Bethlehem' belongs to the Surrey village of Forest Green, for instance. It survives only because of research carried out in the early 1900s.
- The composers Ralph Vaughan Williams and George Butterworth visited isolated villages and wrote down the words and melodies they heard. Ethnographer Cecil Sharp used a wax cylinder, an early sound-recording device. He recorded songs that belonged to particular places and were on the verge of vanishing.
- Some music from these places was later adapted by Vaughan Williams and performed in cities worldwide by orchestras: a splendid example of one place's culture connecting with others globally.

Unique festivals and rituals have been preserved in some rural places in the UK and other European countries (see Chapter 5). Costumed processions, symbolic dramatisations and traditional maypole dances mark the changing seasons. In some places surviving celebrations of the yearly cycle of sowing and harvesting crops have been passed down between generations for centuries or possibly millennia. These local rituals often blend elements of Christian ritual (the harvest festival) with older pagan beliefs (celebrating the cycle of seasons).

▲ **Figure 1.7** The Abbots Bromley Horn Dance and Brockworth's annual cheese-rolling race are both unique rural place-based traditions

- Since 1226, a ceremony called the Horn Dance has taken place in the Staffordshire village of Abbots Bromley. Costumed musicians, including a Hobby Horse and Fool, carry reindeer antlers from the church through the village (see Figure 1.7).
- In the northwest of England, the 'Pace-egging' ritual of begging for eggs is still staged at Eastertime in Heptonstall in the Calder Valley.
- Since the mid-1800s, people from Brockworth, Gloucestershire, have gathered each May to chase a round, locally-made cheese down a hill (see Figure 1.7). Word has spread worldwide through YouTube and the ritual now attracts new contestants from surprising places. A visitor from Japan won the race in 2013; this demonstrates how flows of information and people now connect Brockworth with far more distant other places than in the past.

Place demographics

People are, of course, an important part of the cultural landscape of any place. Over time, different places sometimes develop distinctive demographic characteristics as a result of their local economic functions and site characteristics.

Economic opportunities may attract internal (national) and international migrants. Both migration flows can have a major effect on the age structure, fertility rate and socioeconomic character of a neighbourhood. International migrants additionally bring cultural diversity to a place. In contrast, some places in the UK attract retirees who are far less concerned with the economic opportunities on offer. Instead, these non-economic migrants are more interested in the quality of life gained living in a place with spectacular views or good restaurants and theatres. Table 1.3 explains further reasons why demographic characteristics vary from place to place.

Demographic characteristics	Explanation	Examples
Age profile	■ Population structure may vary from place to place as a result of age-selective migration movements. ■ 'Mover' groups include students, young professionals, couples with young children and retirees. ■ Life expectancy varies markedly between neighbourhoods, according to wealth and poverty levels. As a result, the age–gender structure of places can differ significantly.	■ Some neighbourhoods in Leeds have very high numbers of students as part of a process called 'studentification'. ■ London's Balham neighbourhood is known locally as 'nappy valley', reflecting a high fertility rate due to its popularity with young professional London commuters who are of child-bearing age. ■ Life expectancy for men in affluent Kensington, London, is 84, whereas in Calton, Glasgow, it is just 54 (2016 data).
Socioeconomic profile	■ In-migration of professionals has transformed the population profile of some neighbourhoods. This process is called **gentrification** and it affects both urban and rural areas. It brings rising property prices and changes to the character of local shops and services. ■ In some cities, particularly in the USA, professionals and families have abandoned some inner-city places altogether due to high crime rates.	■ In Liverpool's inner-city Bootle district (Linacre ward), more than 50 per cent of people face significant economic challenges according to local government data. In contrast, in the fringe village of Formby, just 10 km north, less than 10 per cent of people belong to this group while up to 70 per cent are working or retired professionals, including teachers, doctors, secretaries and solicitors. Many have migrated to Formby from other parts of the region.
Cultural (ethnic and/or religious) diversity	■ The ethnic composition of a country, city or local place can change rapidly as a result of the arrival of international migrants and also high fertility among young migrant populations. ■ Cultural diversity can be measured in varying ways, including diversity of nationality, language, religion or race. Migrants may differ from established residents in all or just one of these ways.	■ Over time, some places in London have developed a strong association with Jewish, Muslim or Sikh communities. Sometimes, places of worship or specialist food shops help to 'anchor' some **diaspora** groups to particular places. As a result, some ethnic minority populations in the UK show a high level of segregation in statistical data (see Figure 1.8). Accordingly, some urban neighbourhoods have a distinctive ethnic character.

▲ **Table 1.3** How places may vary according to demographic characteristics and processes

 KEY TERMS

Gentrification The movement of middle-income and high-income groups into places that were previously working-class urban or rural neighbourhoods

Diaspora People with the same ethnic or national roots who live in a range of different countries, such as global citizens of Irish or Indian descent.

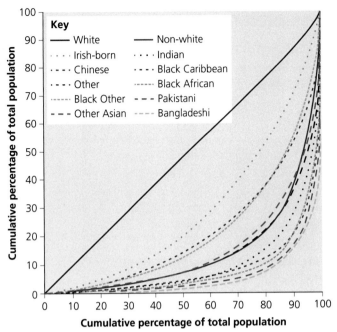

▲ **Figure 1.8** Lorenz curves showing the historic level of segregation for different ethnic communities in the UK in the 1991 Census. Source: Ceri Peach, published in Transactions of the Institute of British Geographers 21(1) p220 (c) RGS-IBG. How might these patterns have changed since, and why?

Places and their boundaries

Finally, one important point to note in this introduction to the study of places is that they may lack clearly identifiable boundaries. In some places, rivers and coastlines provide at least one well-defined settlement margin, whereas local boroughs, wards and electoral constituencies are clearly demarcated on administrative maps. Other places are far less clearly bounded, however. Some urban neighbourhoods do not actually correspond with 'official' administrative areas. In London, it is not clear where Clapham or Chelsea actually begin and end, for instance. In rural regions, topography and vegetation can help create a sense of place, but it is often difficult to establish practically where one upland environment ends and a lowland area begins (see also pages 202–203).

The geographical challenge of boundary-making is indicated by the use of a dashed line in Figure 1.5.

ANALYSIS AND INTERPRETATION

Figure 1.9 shows actual and projected population changes by age group in different UK cities for the period 1993–2020.

(a) For the period 2013–20, identify one demographic characteristic shared by all the cities shown.

(b) For the period 1993–2013, describe how changes in the 25–29 group vary from city to city.

(c) Suggest reasons for the varied changes you have described.

GUIDANCE

Some cities show an increase in the size of the 25–29 group while others show a decrease. Question (c) is best answered by outlining the uneven economic prosperity of different cities during the 1990s following widespread decline in traditional industries in the 1970s and 1980s. London developed a more diverse post-industrial economy more rapidly than many northern cities, in part due to its global influence. London's economic pull factors attracted non-skilled and skilled migrants alike from other parts of the UK. A key element of this 'north–south divide' is the way that demographic changes in different cities are interconnected through migration.

(d) Explain why the changes shown will not occur in every place within each city.

GUIDANCE

Local places within cities do not necessarily experience the exact changes shown at the larger city scale. Inner-city areas that suffered the greatest losses in traditional employment may have experienced even greater out-migration than the whole-city data show. Some affluent neighbourhoods at the rural–urban fringe or in gentrified central areas of northern cities will have gained 25–29-year-olds, particularly towards the end of this time period. Equally, trends may have been uneven within London (some inner-urban areas were still experiencing decline in the 1990s). A well-supported explanation ought to provide named examples of actual places within some of the cities shown.

Population change by age group and region

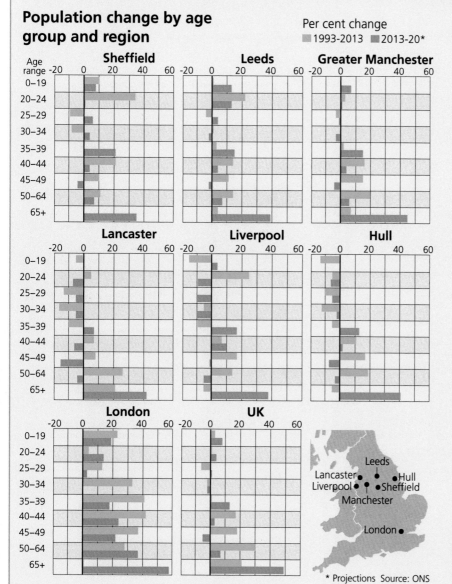

▲ **Figure 1.9** Actual and projected population changes for major UK cities, 1993–2020

* Projections Source: ONS

Pearson Edexcel

AQA

OCR

WJEC/Eduqas

CONTEMPORARY CASE STUDY: HAMPSTEAD HEATH

London's famous Hampstead Heath has a naturally advantageous physical site and is circled by high-class neighbourhoods (see Figure 1.10). Originally several kilometres distant from the edge of the medieval city, Hampstead Heath is around 100 m higher than the city of London. This raised elevation is a result of ancient sea-level changes. The River Thames has eroded downwards into its original floodplain, leaving relict river terraces to the north and south of London. These strips of high ground are identified easily on Ordnance Survey maps.

The prestigious 'village' neighbourhoods of Hampstead and Highgate are sited adjacent to the heath's common lands. These places have housed affluent Londoners ever since they were absorbed by London's urban sprawl in the nineteenth century. During the Industrial Revolution, they enjoyed two important advantages. First, the smog which choked central London rarely bothered the residents of high-altitude Hampstead or Highgate. Second, these places had their own supply of spring water and so escaped the water-borne cholera epidemics of the 1800s.

With spectacular views over London, it is easy to see why these neighbourhoods were originally favoured by the aristocracy and the upper classes. Today, house prices near Hampstead Heath remain well beyond the reach of ordinary people. A five-bedroom house with a view of the heath cost up to £20 million in 2017. This has impacted on social and cultural characteristics and diversity.

The population profile is relatively elderly: life expectancy is 84 compared with the national average of 78 (a reflection, perhaps, of the elite status of many residents and their access to excellent private healthcare).

In the 2011 Census, 84 per cent of Highgate's population was white, compared with 76 per cent for London as a whole.

High levels of education and political influence among Hampstead residents may help explain the area's reputation as a stronghold of **NIMBYism**. The Highgate Society is a pressure group which tries to protect local place character by minimising evidence of globalisation. Legal action by local residents prevented McDonald's from moving into Hampstead High Street for many years. Permission was finally granted in 1992 following years of dispute and only after McDonald's agreed to adopt a very plain shop front.

However, more global high-street brands have begun to arrive in recent years. The local population is changing too as a result of global connections. Overseas property buyers have flocked to London since the 1990s: Russian, Chinese, French and Middle Eastern money has poured into the Hampstead and Highgate property market. The 2011 Census showed a marked rise in foreign nationals living there.

▲ **Figure 1.10** The view from Parliament Hill on Hampstead Heath

 KEY TERMS

NIMBYism The occurrence of 'not in my back yard' attitudes. Local people may, in theory, support a new development such as wind turbines, but they want them built in a place other than their own.

Place dynamics

▶ *How are dynamic internal and external forces responsible for changing place characteristics?*

Changing places, changing connections

Changes over time in the character of local places are the result of two sets of changes. The first set consists of endogenous factors affecting any of the place elements shown in Figure 1.5. Buildings and other structures may be constructed, demolished or replaced for many reasons which are independent of any outside influence, for instance. The second set of causes are exogenous factors: a particular place becomes altered on account of its connections with other near or far places. Places and societies across the entire world have been affected by China's economic transition since the 1980s, for instance. This most likely includes the place where you live:

- Do large, flat-screen televisions made in China hang visibly on the wall in the front rooms of houses on your street?
- Do some of your neighbours light up the exterior of their houses at Christmas with Chinese-made lights?
- How have your local shops and businesses been affected by sales of cheap Chinese goods by online retailers such as Amazon?

These are subtle changes, however. In contrast, some places in the UK have changed beyond all recognition in recent decades on account of exogenous forces. In extreme cases, such as London's Docklands district, a complete transformation has taken place. Globalisation and the global shift of traditional industries has led to the redevelopment of many post-industrial urban places (see Chapters 3 and 4). Work in labour-intensive factories has all but disappeared in many settlements and been replaced by 'white-collar' work in services and consumer industries. In turn, it has become necessary to redevelop many older retailing and service areas in order for them to maintain a competitive edge over rival places. A 'before and after' comparison of the land in and around Birmingham New Street station shows how wholesale replacement of the urban fabric has taken place in recent years (see Figure 1.11).

This section of Chapter 1 also explores the growth of the post-productive countryside, and the implications of rural change for villages and market towns whose economies have been restructured in parallel with the evolution of post-industrial urban areas. Rural places in the UK are no longer characterised mainly by agricultural land uses and employment. Rural diversification has brought a broader mix of tourism, sports, leisure services and artisan crafts to the British countryside.

 KEY TERMS

Endogenous factors Place-making factors that originate internally. They might include aspects of a site or land on which the place is built such as the height, relief and drainage, availability of water, soil quality and other resources. They also include the pre-existing demographic and economic characteristics of the area, as well as aspects of the built environment and infrastructure.

Exogenous factors Place-making factors that originate externally. They include links with, and influences from, other places. These relationships with other places may include the movement or flow of different things across space such as people, resources, money, investment and ideas.

Global shift The international relocation of different types of economic activity, especially manufacturing industries. The term is strongly linked with the writing of Peter Dicken.

Rural diversification The evolution of rural economies to include a much wider range of activities than agriculture only. Tourism and rural crafts are two important additions.

▲ **Figure 1.11** The space in and around Birmingham New Street station has been radically redeveloped between 1990 (above) and 2015 (below)

Endogenous (internal) factors

The physical geography and landscape of a place may change over time, irrespective of any external connections with other societies. Places are constantly modified by people living there in deliberate ways. A good example of this is land reclamation alongside rivers giving rise to new kinds of land use. The construction of London's Victoria Embankment in the 1860s narrowed the River Thames. In a feat of engineering, marshy river banks were replaced by streets and parks; new sewers and an underground train line were installed below.

A location may undergo change because of environmental factors. Sometimes, the physical geography that favours a settlement's growth plays a role in its later downfall. When non-renewable resources such as coal, copper and tin become exhausted, productive industries cease: mining 'ghost towns' are found the world over, from Chile to Australia. When renewable resources like soil and vegetation are used in unsustainable ways, places sometimes suffer decline too. Farms in Oklahoma and Texas were abandoned during the USA's 'dustbowl' episode of the 1930s. Strong winds had combined with unwise farming practices to strip the soil from the land.

🔑 **KEY TERM**

Deindustrialisation A decrease in the importance of industrial activity in a local place or wider region, measured in terms of employment and/or output.

The city of Chester underwent **deindustrialisation** centuries ago (long before global shift). The naturally-occurring process of silt deposition in the Dee estuary led to fewer ships being able to reach Chester by the 1700s. Eventually, the city lost its shipping trade and port industries to neighbouring Liverpool and the deeper waters of the Mersey estuary. Chester subsequently changed as a place: a new economic and cultural landscape developed that was focused less on manufacturing industry and more on shopping and services.

Deindustrialisation may also occur when a society's perception of local environmental risk changes. Improved understanding of the dangers associated with lead and asbestos mining helps explain the decline of these activities in the UK and a similar process partly accounts for the global shift of coal production. Workers' trade unions fought successfully to modernise British coal mines and insisted on protective equipment and clothing as the health risks became clearer. Now it is cheaper to import coal from countries where low-paid miners still suffer accidents routinely, such as South Africa, and production has all but vanished from the UK (see pages 82–83 and 94).

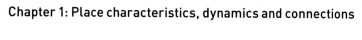

The pattern is repeated for many other primary, processing and manufacturing industries. Dangerous or polluting industries have become financially unviable in the UK due to exhausted reserves (the most accessible and cheapest were extracted first), stricter laws, trade union pressure and corporate social responsibility.

Exogenous (external) factors

Place changes are also driven by external events, issues or processes. Some of the endogenous factors outlined above have exogenous dimensions which relate to globalisation, a process which has brought economic and cultural change to rural and urban places throughout the UK. The effects of global flows of people, money, technology and information are clearly visible in local high streets everywhere. Inescapable macro-scale forces have restructured the British economy and people's work. Diversity of food, fashion and faces has increased in smaller towns and rural places, not just in major cities.

Viewed historically, however, globalisation's effects are merely the latest round in a far longer series of externally-driven changes influencing the characteristics of places:

- *International migration* from distant places in mainland Europe brought all kinds of demographic, economic and cultural changes to early settlements in the British Isles. Roman, Viking and Norman invaders all left their mark (think of the origin of the names we use for days of the week). More recently, **post-colonial migration** and the free movement of workers from the European Union have combined to bring unprecedented cultural diversity to some urban neighbourhoods. An estimated 300 languages are now spoken in parts of London.

- *Global and national resource flows* fostered the industrialisation of urban places in the 1700s. World trade and the British Empire supplied Sheffield's steel industry with ivory and bones for its knife handles; and some places and groups of people living in Sheffield grew very wealthy from exporting cutlery to national and international markets. In past centuries, British citizens travelled widely overseas before returning with new foods, materials or ideas. Potatoes, tea, coffee and tobacco were all brought back to the UK by early explorers. Imagine how different British high streets would be without chip shops and cafés! Consider also the enormous harm that smoking brought to communities throughout the UK during the heyday of this damaging habit. In all places, life expectancy was reduced when the majority of people smoked.

- *Rural–urban migration* transformed urban societies and landscapes during the late 1700s and 1800s. Every large town and city is the result of an influx of rural migrants experienced at some juncture in that settlement's history. The importance of this process cannot be over-stated. Across the world,

KEY TERM

Post-colonial migration
People moving to the UK from former colonies of the British Empire during the 1950s, 1960s and 1970s. Alternatively, this movement is sometimes described as the arrival of 'the Windrush generation'.

▲ **Figure 1.12** The eighteenth-century cartoonist Hogarth portrayed the horrors caused by alcoholism in a squalid London place he called 'Gin Lane'

🔑 **KEY TERM**

Time–space convergence The lived experience of distant places feeling nearer due to the 'annihilation' of distance by new transport and communications technologies.

rural people are an essential ingredient for the engine of urban growth, be it in nineteenth-century Manchester or twenty-first-century Lagos. In the UK's urban areas, soaring population and housing density growth during the Industrial Revolution at first led to widespread squalor and congestion. Urban neighbourhoods suffered from overcrowding, poverty, alcoholism and epidemics of disease (see Figure 1.12). Later, Victorian engineers tackled these problems with innovative new infrastructure, including public transport and sewer systems, thereby changing urban places in highly disruptive but ultimately positive ways.

New technology and flows of ideas

Technological changes have affected places in both positive and negative ways for millennia. Imported technology and new ideas sometimes help places grow in new and changed ways, but the reverse can happen too. Technology may make places obsolete; this is a logical consequence of time–space convergence. This process was described by Donald Janelle in 1968 as the 'annihilation' of distance by successive waves of transport innovation. When railways replaced stagecoaches and canals in England, distant places were brought closer together in 'time-space' (see Figure 1.13). Some settlements suffered as a result, however, when the canals they were built alongside fell into disuse. The growth and later decline of the Yorkshire mill town of Todmorden is partly explained by the changing importance of the Rochdale canal between the 1700s and 1900s.

Time–space convergence has been responsible for the rise and fall of many places in history. Another example is the increased size of container ships used during the 1960s. These new vessels were too large for the River Thames and places in London's Docklands suffered disintegrating economies when river trade ended abruptly (see pages 158–159). In the Isle of Dogs and Limehouse, male unemployment had reached 60 per cent by the end of the 1970s. For every job lost at the docks, three more disappeared in support industries, including transportation and warehousing. Local pubs and cafés lost trade too. In total, almost three-quarters of a million jobs vanished from the capital between 1960 and 1975. The city's population reduced in size from its 8.6 million peak in 1939 to 6.7 million by 1981.

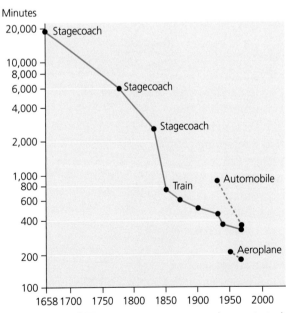

▲ **Figure 1.13** Time–space convergence demonstrated by the reduction in travel time between Edinburgh and London, 1658–1966

External forces acting on rural areas

Endogenous factors played a major role in ending the productivist phase of the British rural economy. The countryside was traditionally a space of production where communities grew crops and reared livestock that they either consumed or traded in urban markets. Lancashire was known for dairy farming, East Anglia for cereals and so on. But by the 1970s, this way of life was ending, often for reasons similar to those underpinning the loss of manufacturing industry in British cities.

Globalisation and the enlarged EU free market both meant cheaper imports of meat, cereals and fruits arriving in the UK. Reduced state subsidies for farmers further undermined the economic stability of some rural places (though for many farmers this was balanced by increased EU spending on agriculture). The countryside had also benefited in the immediate postwar decades from generous regional development policies for 'problem regions' worth hundreds of millions of pounds. However, substantial cutbacks in public spending followed the high-inflation period of the mid-1970s.

These mounting pressures further accelerated the mechanisation of farming and the use of arable contractors instead of permanent employees on farms. Combine harvesters were first introduced after the Second World War when the UK Government used US financial aid to boost domestic food security.

- By the 1970s, farm machines were a common sight but at the cost of yet more jobs. In 2015, just over one per cent of the British workforce worked in agriculture compared with eleven per cent a century earlier.
- Many small villages lost large numbers of people after 1945 as farming employment dwindled. Some were subsequently repopulated by city commuters and second homeowners (see pages 193–195).

It is important not to overstate the decline of agriculture *as land use*, however. Even today, around 70 per cent of UK land is used for agriculture. In reality, the post-productive countryside is made up of places where new forms of work are often becoming available *alongside* mechanised farming. Chapter 6 explores how rural places have diversified their economies by embracing consumer-orientated leisure and tourism services. Hospitality and retailing have replaced agriculture as the main employers in much of the British countryside.

▼

ANALYSIS AND INTERPRETATION

Figure 1.14 shows the number of London properties bought by overseas companies between 2010 and 2015. Also shown are the territories where the main buyers are based.

(a) Describe the distribution of London properties bought by overseas companies.

(b) State one possible change you might make to the map and explain why this would be an improvement.

GUIDANCE

In addition to using a variety of skills and techniques, Geography students should also be able to reflect critically on the methods they use. The diagram shows a choropleth map. The data have been divided into five categories; the number of properties bought in each local authority area is shown (rather than smaller administrative sub-areas). Do you agree with both of these decisions? Could the map be improved and, if so, how?

(c) Suggest what Figure 1.14 shows about the way place characteristics can be influenced by connections with other places.

GUIDANCE

The diagram shows that properties are being bought by overseas buyers, suggesting large inflows of capital into the London housing market. What further effects could this trend have? The current unaffordability of housing is linked with strong demand among overseas buyers and the price increase this causes. Where demand is focused in particular postcodes, local prices are driven to especially high levels. How might local places be changed if overseas buyers decide to live in these properties rather than rent them? Could local demographic characteristics begin to change? Would local shops and restaurants change their services and menus to cater for high-income migrant communities?

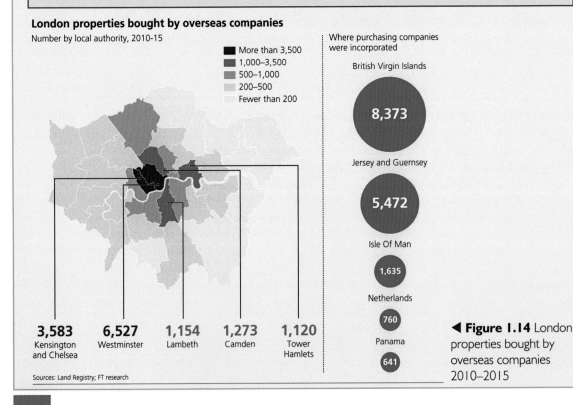

London properties bought by overseas companies
Number by local authority, 2010-15

More than 3,500
1,000–3,500
500–1,000
200–500
Fewer than 200

Where purchasing companies were incorporated

British Virgin Islands

8,373

Jersey and Guernsey

5,472

Isle Of Man

1,635

Netherlands

760

Panama

641

3,583	6,527	1,154	1,273	1,120
Kensington and Chelsea	Westminster	Lambeth	Camden	Tower Hamlets

Sources: Land Registry; FT research

◄ **Figure 1.14** London properties bought by overseas companies 2010–2015

CONTEMPORARY CASE STUDY: INNER-CITY SHEFFIELD

Local and external factors have combined to shape the economies and identities of places in both positive and negative ways over time. Take the case of Sheffield and the surrounding villages of Hallamshire. Locally-sourced iron and millstone grit were being used to make knives and scythes there by the thirteenth century. Demand from other places in England help explain the region's very early success. By the late 1800s, Sheffield had become a city serving global markets with mass-produced outputs of steel and cutlery. In turn, the city's people now shared a strong sense of identity. Workers from Sheffield's metal factories and workshops founded a football team nicknamed 'The Blades', for instance (see Figure 1.6).

This confidence did not last. International demand for Sheffield steel and metal fell sharply in the 1970s. Global shifts meant South Korean steel-makers had captured the lion's share of world markets. Inner-city places in Sheffield, like the Park Hill and Manor estates, suffered more than a catastrophic economic slump; they experienced an identity crisis too. Neighbourhood life was, for most people, no longer based around the shared experience of making steel and metal.

Individual places in Sheffield like Park Hill have their own unique histories of course. The Park Hill site was once a deer park for the Earl of Shrewsbury. It became an area of tenement housing during the Industrial Revolution, known as 'Little Chicago' on account of its high crime rates in the 1930s. During the 1950s it became the site of a new housing complex (see Figure 1.15). Influenced by imported architectural designs and ideas from continental Europe, the Park Hill estate has subsequently gained Grade II* listed status.

Today, there are few visible clues left in Sheffield's skyline to suggest its heavy industrial past. However, newer forms of industry are actually thriving in certain places, sometimes housed within the shells of old factories. For instance, the university's Advanced Manufacturing Research Centre (AMRC) employs 500 engineers and scientists to make specialised titanium and tantalum products. Steel may no longer be manufactured there but the city has entered a 'third age' of metal working for international markets. Pre-industrial metal craftwork preceded the second industrial phase of mass-produced steel and cutlery; this gave way to **post-Fordist activity**. But without the region's physical geography, and the metal-working and engineering skills that developed *in situ*, none of it would have been possible.

▲ **Figure 1.15** The Park Hill estate, Sheffield

🔑 **KEY TERM**

Post-Fordist activity Flexible and market-driven work which is highly responsive to consumer demand, especially for luxury goods. Some post-Fordist industries specialise in the production of luxury or specialist goods, often made by hand using traditional techniques. 'High tech' industries, including robotics, precision engineering, jet engines and weaponry makers, are another category of post-Fordist industries.

3 Place networks and layered connections

▶ *How does the study of place networks and 'layered connections' help us to understand place identity?*

Past and present networks of places

Different localities become linked together to form a **network** of connected places, both in the past and the present. A network is an illustration or model that shows how different places are linked together by connections or flows, such as **foreign direct investment (FDI)** made by transnational corporations (TNCs) or migration. Network mapping differs from topographical mapping by not representing real distances but instead focusing on the varied level of interconnectivity for different places – or nodes – positioned on the network map (see Figure 1.16). Especially well-connected cities and particular places within them – such as university research clusters in Cambridge or Manchester – are described as **global hubs** in network theory. Key network flows comprise:

- *Capital (money)*. Important financial flows that connect some places with distant places elsewhere include TNC investment in commercial districts and residential purchasing by overseas investors (see pages 128–129).
- *Raw materials*. Commodity flows include crops, minerals and fossil fuels, channelled not only by the operations of major companies but also by small businesses and individuals trading online. Surprising commodity connections sometimes exist between places. Calder Textiles in the Yorkshire town of Dewsbury makes hand-dyed cashmere yarn for fashion houses. It sources the wool from goat farmers in Afghanistan's Herat province. This partnership between places was brokered by the US defence department!
- *Merchandise*. Around 200 million container movements take place each year on sea-faring vessels. China – the source of so many merchandise flows – is often called the 'workshop of the world' (an accolade belonging to the UK in an earlier era). Bangladesh and Indonesia are also important producer nations. Many UK high streets have a discount store where cheap goods sourced from these countries are sold in large volumes.
- *Information*. Places everywhere are increasingly exposed to data flows. Radio, television and often broadband are available in the remotest parts of the UK. Elsewhere in the world, remote rural communities in India have mobile phone services. So too does the tiny Inuit village of Little Diomede, which is located on an isolated island in the Bering Strait, with Alaska to the east and the Russia to the west (this is despite the fact that the mail only arrives once a week here).

KEY TERMS

Network This is an illustration or model that shows how different places are linked together. Geographers create networks to emphasise the connections that exist between different places.

Foreign direct investment (FDI) A financial investment made by a TNC or other international player (such as a sovereign wealth fund) into a state's economy.

Global hub A settlement (or wider region) providing a focal point for activities that have a global influence. All megacities (10 million or more people) are global hubs, along with some smaller settlements such as Cambridge, whose university and science park have a truly global reach.

- *Services*. Cornwall and the Highlands of Scotland are regions where a growing number of self-employed people offer 'teleworking' services. Architects, solicitors, writers and software designers work remotely from their home offices. Their clients may live anywhere.
- *People*. In 2017, an estimated 250 million people were living in countries they were not born in, up from 80 million in 1965. This figure includes legal and illegal economic migrants (over 150 and 50 million respectively), as well as displaced persons (refugees escaping from physical disasters or political persecution). Even some of the UK's most rural remote places have experienced in-migration, including the Rhondda Valley in Wales, where several Indian-born NHS doctors and their families have lived for many decades.

Improvements in both the speed and capacity of transport and ICT (information and communications technology) are frequently cast as the key 'driver' in histories of place network growth at varying scales. Important developments of the last thirty years – the internet, mobile phones, low-cost airlines – have certainly increased place connectivity. Economic transactions are easier to complete, be they productive (manufacturers can contract and outsource physical goods from increasingly far-away places utilising ever-faster transport networks) or consumerist (goods, information and shares can seemingly be bought anywhere, any time, online at the touch of a button). People move about more easily than in the past too. Budget airlines bring a 'pleasure periphery' of distant places within easy reach for the moneyed tourists from high-income nations.

(a) Places shown as territories on a map

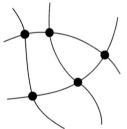

(b) Places shown as nodes in a network

▲ **Figure 1.16** Places in (a) topographical and (b) network mapping

A shrinking world and changing perceptions of 'near' and 'far' places

Heightened connectivity changes our conception of time, distance and potential barriers to the migration of people, goods, money and information. Distant places feel closer than in the past. As a result, the definition of what constitutes a 'near' or 'far' place changes in line with shifting perceptions of spatial relations. Towns in the south of France may feel like 'near' places for English easyJet passengers. A century ago, in contrast, English people would have viewed the Mediterranean coastline as a distant location. According to Janelle's theory of time–space convergence (see page 22), technology makes places feel closer together than in the past. Each successively improved transport technology chips away more minutes and hours from the connecting journey's duration. Since the sails of ships first filled with air, human society has experienced a 'shrinking world'.

Writing in a similar vein, David Harvey has argued that these time–space changes have been crucial for the survival of capitalism. Technology has been pressed into service by global economic empire-builders. Fast trains and broadband connections – like the aeroplanes and container ships which move commodities around the world – are an outcome of the never-ending search for new markets and profits by TNCs.

A second axis of human power – the exercise of military might and imperial ambition – also continues to provide an equally important stimulus for new transport and communications innovation. One of the earliest shrinking-world technologies clearly served nations' security: the practice of lighting warning fires across a chain of beacon hills dates back to ancient Greece. The first sighting of the 1588 Spanish Armada was also signalled across England in hours by hill-top bonfires. The people responsible for lighting them created an early network of connected places.

One important caveat must be added to the study of connected places in a shrinking world, however. Doreen Massey was a geographer who wrote critically about changing perceptions of place in a technologically advancing world. She argued that time–space convergence is socially differentiated: not everyone experiences the sense of a shrinking world to anything like the same extent. Today, privileged elite groups routinely fly around the world for reasons of work and leisure: academics attend international conferences and rock stars enjoy their stadium world tours. In contrast, many more people's lives have been transformed by technology only in so far as they now consume a glut of cheap imported food, goods and television shows.

Multi-layered connections

Place stories usually have several chapters because of the way place networks and flows have changed over time. Each historical period has its own connections with other near or far places (see Figure 1.18). Over time, layers of connections built up to produce an accumulated history that is visible in each place's cultural landscape. Surviving buildings from many eras can be seen in York, including the famous medieval Shambles (see Figure 1.17), while Roman pottery and Viking remains have been unearthed from beneath the city's streets by archaeologists. These historical artefacts reveal how international flows of goods, people and money have shaped these places over millennia to produce a 'nexus' of connections and linkages. The previous case study of Sheffield (see page 25) also illustrates how linkages between a settlement and the wider world build up, layer upon layer, over time.

▲ **Figure 1.17** The Shambles in York

Connections can affect places in both positive and negative ways (and perspectives may vary on which is which).

- Recent terror attacks – related in part to complex historical and political connections between the UK and other parts of the world, including the Middle East – have had a profound effect on urban spaces and places in the UK. In particular, the 'weaponisation' of vehicles has meant pavement barriers have been introduced to vulnerable places, for instance along Westminster Bridge. New plans to pedestrianise Oxford Street in London are concerned not only with air pollution but will help deny access for potentially weaponised vehicles to an intensively used recreational space (along with heightened CCTV surveillance, this is sometimes called the 'securitisation' of public space).
- Liverpool Football Club became embroiled in a backlash against China's alleged human rights abuses in Tibet after campaigners called for it to cancel a sponsorship deal with a company that bottles water at a Tibetan glacier. As European football clubs attempt to reach out to investors in China, some of their supporters are opposed to the increase of place connections they deem unethical.
- There is growing public criticism of well-connected companies based in the UK moving their profits offshore to avoid paying tax.

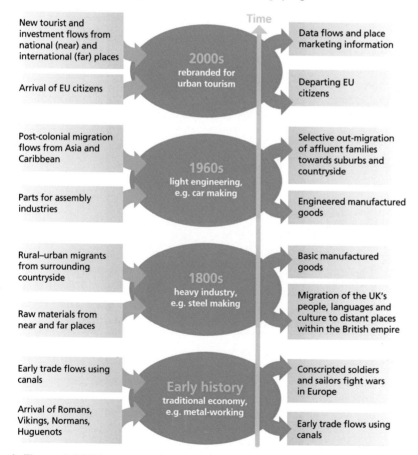

▲ **Figure 1.18** 'Time-space' changes in an inner-city neighbourhood: 'layered connections' have developed with other near and far places. Local- and international-scale connections feature in this diagram: can you identify them?

CONTEMPORARY CASE STUDY: BARCELONA

The majority of examples and contemporary case studies used in this book are drawn from the United Kingdom. However, if your A-level specification requires you at certain points to study 'contrasting' places then it could be a good idea to research a non-UK-based case study.

The Spanish city of Barcelona features on the cover of this book and contains many smaller-scale places worthy of investigation. This settlement's rich history can be used to illustrate many changing places themes, including urban planning, rebranding and regeneration ('place remaking') and the role of global businesses and technology.

The Eixample is a district of Barcelona designed by Ildefons Cerdà and constructed in the nineteenth and early twentieth century. It is an ambitious exercise in urban planning and a world-famous example of 'utopian urbanism' (see page 59). This place's highly distinctive octagonal blocks, with chamfered corners, are shown in the cover photograph.

Barcelona is a widely-cited example of urban redevelopment. The staging of the 1992 Olympic games was a watershed moment for the city and paved the way for city-wide redevelopment including widespread slum clearances in a place called El Raval, the historic district adjacent to the original Roman and medieval city, the Barri Gotic. Some methods were controversial, including the demolition of rundown buildings and 'decanting' (displacement) of the original population to other areas. Chapters 4 and 5 explore similar themes in UK contexts.

The annual number of overnight visitors has grown from around 2 million in 1990 to over 8 million today and the number of hotel beds has tripled. Tourism accounts for fourteen per cent of the economy and provides over 100,000 jobs. But in recent years, tensions have grown between residents and holidaymakers. When newspapers around the world ran photos of naked Italian tourists cavorting through one Barcelona neighbourhood in the middle of the day, widespread protests were staged against out-of-control tourism.

In parallel to its growth as a global tourist hub, the city has hosted numerous global technology fairs and in 2014 was awarded the title of European Capital of Innovation by the EU. The city authorities have promoted Barcelona as a global-class 'smart city' (see page 137).

Barcelona has a complex identity because it is embedded in the culturally distinctive region of Catalonia. Many Barcelonan citizens seek full independence from Spain and Catalonian flags are a common sight in the city. Support for independence has climbed to more than 50 per cent in recent years.

ANALYSIS AND INTERPRETATION

In the following qualitative data source, central London is portrayed as a dynamic place whose landscape is filled with clues about past and present geographical connections with other places in the UK and beyond. Read the passage and answer the questions.

◀ **Figure 1.19** London under construction

Construction in London

In London, most urban construction today is perceived as a nuisance rather than a sign of progress, greeted with frustration: a closed Tube station, a road diversion, noise. As the capital grows, it goes through waves of rebuilding, each supposedly addressing a dominant issue. In the late nineteenth century it was slum clearance; after the Second World War it was the rebuilding of a city devastated by bombing; in the 1980s the rebuilding was an effort to revitalise the city as a global financial centre. Today, housing supply needs to keep up with population growth.

The biggest holes that have appeared in central London during the past few years have been those dug to build the £15 billion Crossrail project, which, by the end of 2019, will provide a 100 km rail link from Reading and Heathrow in the west, under central London, to Shenfield in the east. Chasms have opened up in some of the city's most tightly knit and atmospheric neighbourhoods. Big chunks of Soho have been concreted over. The blocks around Denmark Street, once the hub of the music business with its guitar shops, recording studios and basement clubs, have been fast disappearing. This messy neighbourhood, so dirty and incoherent it was overlooked by developers, is now being repackaged as a piece of prime real estate with added musical nostalgia.

There can, of course, be no construction in the city without demolition. The sudden exposure of a site leaves the surrounding buildings looking vulnerable. It gives a glimpse of the parts of structures never meant to be seen from the street. It opens strange new views that can leave the onlooker surprised, even disoriented. Exposed interiors can be seen, often complete with wallpaper and fireplaces stuck halfway up walls. Period façades are retained in a futile gesture to maintain the appearance of the historic city while most of the rear is replaced with steroidal new development.

But one thing these demolition gaps also reveal is the astonishing complexity and *ad hoc* construction of fragments of a city that have been added to and adapted over centuries. Often the original bones are impossible to detect, obscured by decades of accretions. These were almost always buildings with elegantly clothed façades; the architecture was only for the front. Yet these Georgian and Victorian creations have managed to survive through the centuries, accommodating adaptation and reuse almost in every decade; from residential to commercial, workshops to studios, shops, bars and bedsits as needed.

Source: Heathcote, E. (2016) Construction in London. *Financial Times* [online].

(a) Using the first paragraph, explain how place changes can be caused by exogenous factors.

GUIDANCE

It is a requirement of the question that evidence from the first paragraph of the passage is used as part of the answer. A range of reasons are provided for 'waves of rebuilding' affecting London. The aim of this question is to provide students with an opportunity to apply their knowledge of what is meant by 'exogenous factors' to the context provided.

(b) (i) Using evidence from the passage, identify ways in which economic functions have changed over time in the city of London.
     (ii) Using your own knowledge, suggest possible reasons for these changes.

GUIDANCE

Mention is made of increasing and decreasing economic activities. Using your own knowledge, you may be able to link the changes the writer describes with broader trends you have learned about, such as deindustrialisation or globalisation.

(c) Using evidence from the passage and your own knowledge, explain how the characteristics of places are shaped by past and present connections with other places.

GUIDANCE

This question invites an examination of the multi-layered complexity of places. A good approach might be to contrast the example provided of London with a parallel account of your own home place, city or region.

④ Evaluating the issue

▶ *To what extent can places be completely protected from change?*

Possible contexts for protection and change

Early-industrialising countries like the UK, France and the USA have experienced centuries of economic growth and rural–urban migration. During this time, settlements have been affected by important functional and landscape changes as their economies have matured. Additionally, some landscapes contain archaeological evidence of more ancient settlement and civilisation pre-dating the Industrial Revolution.

- Older places are often characterised by a 'palimpsest' landscape. This means it contains traces of both recent and far older societies, activities and cultures. In successive historical periods, new layers of development were laid down like a blanket (Figure 1.20). But these blankets contained holes and did not cover *everything* that had come before. Roman, Norman and medieval churches and houses can be seen in many older British towns and cities, despite being surrounded by more recent housing and offices.
- The oldest surviving structures are generally those with special uses and meanings attached to them, such as cathedrals, museums and town halls. In recent decades, however, a wider range of building types has been valued and protected, including brutalist tower blocks and other forms of modern architecture.

For comparative purposes, it is worth considering the extent to which these patterns and trends are also true for cities in developing countries and emerging economies.

- Many cities in Asia, Africa and Latin America have ancient histories. Since 1945, many important surviving buildings and places have been awarded World Heritage Site (WHS) status by the United Nations Educational, Scientific and Cultural Organization (UNESCO), whose aim is the 'preservation and promotion of the common heritage of humanity'. Cairo's Old Town, where

▲ **Figure 1.20** Canals and warehouses from the early 1800s still show through the gaps in layers of more recent urban development at Victoria Quays in Sheffield; Park Hill (page 25) is visible on the horizon

Sultan Hassan Mosque is situated, has WHS status.

- However, population pressure on Cairo and other developing world cities is even greater than that experienced by Europe in the past. Some megacities like São Paulo and Lagos now receive half a million new arrivals annually. Shenzhen in China has grown from a small market and fishing town of fewer than 300,000 people to a sprawling metropolis of 20 million people in just 35 years. Growth is so rapid that protection of older districts becomes a challenge.

In contrast to cities, rural areas are sometimes relatively easy to insulate from change. In Europe, the countryside is often highly protected, especially in places where ancient monuments and architecture are found. Legal restrictions limit new development on UK greenfield sites. There is less demographic and commercial pressure to develop isolated rural places in any case.

- Rural land ownership plays an important role: Scotland's Isle of Jura has been preserved and protected by its owners. It functions as a hunting estate where land use changes have been kept to a minimum.
- However, unpreventable changes to the physical environment, such as those associated with climate change, may be far more clearly visible in a rural rather than an urban context.

Evaluating the view that places *can* be protected

There are many examples of protected places. Ultimately, most land markets are highly regulated. Planning laws and rules control the changes that may occur legally. In UK towns, any proposed land-use alteration (from residential to commercial use, or vice versa, for instance) is considered carefully by the local authority. Many applications are rejected. Also, over 400,000 buildings in the UK now have Grade I, II* or II listed status, which gives them a high level of protection. Since 1983, the government-funded agency English Heritage (now known as Historic England) has overseen this process. All buildings

built before 1700 which survive in anything like their original condition are listed, as are most of those built between 1700 and 1840.

The valuing of landscapes and places can become a controversial and contested process, however. Not everyone will agree on what should be protected and what should not. Inevitably, subjectivity, individual perspectives and bias affect decision making. Tension can result if some people find special meaning in a place while others do not (see Section 1.2).

- The 'top-down' decision by a government or UNESCO to preserve a place may meet with 'bottom-up' resistance from local people who would prefer to see fresh commercial development and new employment opportunities.
- Historic England has added modern buildings to its protected list, including the brutalist Park Hill estate in Sheffield (see page 25), which has architectural merit according to experts. This view is not shared by many people.

Sometimes, it makes commercial sense to protect places rather than develop them, of course. Heritage tourism can be rewarding: the historic Belgian city of Bruges (Figure 1.21) has seen very

▲ **Figure 1.21** Bruges is a globally-connected city on account of its picturesque and protected 'fairy tale' skyline

little modern development and its picturesque skyline draws millions of visitors annually. New developments in Oxford and Bath are closely monitored in case they detract from the historic landscape visitors expect to see. City planners in London have allowed a mixed modern and historic landscape to develop but no large new construction is approved if it blocks the protected view of St Paul's Cathedral from various prominent locations around the city.

- Green belts remain in place around the UK's major cities, despite rising demand for new affordable housing.
- Some places within UK towns and cities have been designated as conservation areas where trees are protected and cannot be cut down without special permission from the local planning authority.

Evaluating the view that places *cannot* be protected

Many counter-arguments can also be made, showing that place protection can, in reality, be hard to achieve. The sky-high value of the land in world cities like New York and London means that large commercial developments can yield enormous profit. The stakes are so high that developers are often prepared to fight long and expensive legal battles to gain planning permission.

Population pressure cannot be ignored either. Migrants are drawn disproportionately to high-performing cities like London and Cardiff. City planners have no control over immigration into the UK. Instead, they must attempt to accommodate population growth as best they can. What options are there other than to grant permission for the construction of more housing, primary schools and supermarkets? External pressures mean that proposals for new developments and housing extensions are more likely to be accepted; 'you can't stop change' becomes the pragmatic and prevailing response.

Additionally, laws and principles of protection will always be breached in exceptional circumstances:

- Little can be done to protect places from harm in times of war. Liverpool and London were damaged greatly by Second World War bombing, while Coventry lost its cathedral during the conflict. More recently, the Roman temple at Palmyra, in Syria, was destroyed by Daesh (or so-called ISIS) forces. They did not care that UNESCO had awarded it WHS status.
- Extreme weather events transform places through the havoc they wreak. In 1976, southern England experienced an eighteen-month drought that killed many trees. A further 15 million English trees were felled by a great storm in 1987. Hurricane Katrina obliterated entire urban neighbourhoods of New Orleans in 2005.

Increasingly, many places are affected by uncontrollable longer-term climate change. Rising sea levels mean that many coastal areas must either be abandoned or acquire new defences. Either way, place change is inevitable. Landscape characteristics of rural areas will gradually be modified as rainfall and temperature patterns change. In southern England, ecosystems like Epping Forest may not survive in their current form if climate change projections are correct.

Arriving at an evidenced conclusion

To what extent can places be completely protected from change? Realistically, total protection – in the long-term – is impossible. Around the world, different places are threatened by unpreventable earthquakes, hurricanes, tsunamis, coastline retreat and climate change. Adaptation through engineering or out-migration may be the only response: inevitably, affected places are changed.

Pearson Edexcel

AQA

OCR

WJEC/Eduqas

Hard to avoid too are globalisation and global developmental processes, including changing fertility and mortality rates, and the growth of social media. Few places are truly switched-off from these trends. Even relatively isolated places and populations change in line with global development over time.

However, place change can potentially be minimised in the short term by effective governance. Enforceable local planning controls and laws are required for NIMBY ('not in my back yard') and 'preservationist' attitudes to prevail. Place changes may also be minimised by political decisions taken at the national level limiting global interactions, such as the introduction of stricter migration controls or higher import tariffs on foreign-made goods. Finally, places can sometimes be restored to their previous state after an unwelcome and unpreventable change has occurred. The ability of some societies and places to manage this outcome is called resilience. The reconstruction of historic buildings and monuments after a fire is one example of this.

Ultimately, though, all societies and environments change over longer timescales. Accordingly, so too do places, given that their identity is shaped by the interplay of dynamic social and environmental forces. In the long term, place changes can be slowed down by protective measures but never prevented entirely. In the short term, the degree of protection which is possible depends on (i) the scale and pace of the force of change, (ii) the political will to resist it and (iii) the cost of doing so.

 KEY TERMS

Brutalism A mid-twentieth-century architectural style typically characterised by massive, plain and unadorned concrete blocks and towers. Brutalist designs are now viewed widely as fierce and aggressive rather than inspiring. The term was originally coined in Sweden in 1950 to describe boring brick building designs.

Emerging economies Countries that have begun to experience higher rates of economic growth, often due to rapid factory expansion and industrialisation. Emerging economies correspond broadly with the World Bank's 'middle-income' group of countries and include China, India, Indonesia, Brazil, Mexico, Nigeria and South Africa.

Greenfield site A place which has remained free of housing, industry or urban development (although it may have been used for farming) and which remains 'natural' in appearance.

Affordable housing New homes which are sold at below-market prices to first-time buyers. Local councils can require new housing developments to include some affordable housing.

Governance The term suggests broader notions of steering or piloting rather than the direct forms of control associated with 'government'.

Resilience This describes the capacity of a place (or society or economy) to return to its previous state, or establish a new growth path, following some major disruption, shock or disaster.

Chapter summary

- Places can be understood as small-scale settlements, neighbourhoods or areas whose physical and human characteristics collectively create a distinctive identity. Site factors, economic functions and the cultural landscape of a place are three interconnected elements which provide a place with its 'personality'.

- The elements of a place are dynamic: internally-driven (endogenous) changes may result from the operation of physical processes or the management decisions of the people living there.

- External (exogenous) connections with other places and people bring frequent changes to places. Migration is an important driver of cultural landscape change on account of how it affects population structure and ethnic diversity. Globalisation is another key influence: the global shift of manufacturing industries had knock-on effects for local economies and societies throughout the UK.

- Rural areas have been changed by links with other places, especially neighbouring urban areas. The migration of city workers into adjacent countryside blurs the distinction between rural and urban places by creating 'rurban' fringe areas or edgelands.

- Geographic connections and network flows change in successive historical eras. Older settlements therefore gain a multi-layered identity which reflects both past and present flows of people, money, resources and ideas.

- No place can be completely protected from change due to the complexity of the many geographic processes and connections which shape the surrounding world. Contingent political and natural events, including wars, extreme weather and climate change, bring destructive and creative effects that cannot always be mitigated against.

Refresher questions

1. What is meant by the following geographical terms? Settlement site; settlement function; cultural landscape.

2. Using examples, outline the interrelationships that exist between the economic function and cultural landscape of different places.

3. Using examples, explain the difference between a rural place and a rural region.

4. Outline different ways of distinguishing between rural places and urban places.

5. Explain the demographic and economic changes which cities in the UK have experienced on account of in-migration from (i) near places and (ii) far places.

6. Using examples, explain two ways in which new technology has brought change to rural places.

7. Outline how different flows of people, resources and ideas create networks of connected places.

8. Using examples, explain what is meant by 'a shrinking world'.

9. Using examples, explain how players at varying geographic scales have attempted to protect different places from change.

Discussion activities

1 In pairs, discuss the characteristics of (a) the place or neighbourhood where you live and (b) the city or region it belongs to. What characteristics does your home place share with the wider city or region? What makes your place unique and distinct from the rest of the city or region?

2 Use Figure 1.5 and Figure 1.18 to help you describe the 'place story' of your home neighbourhood or settlement.

3 Draw a timeline from 1945 up to around the time you were born for (a) the UK as a whole and (b) the place where you live. Add annotations to show any major economic and social changes that you have learned were taking place during each decade of your timeline.

4 In pairs, make a list of as many of the UK's largest towns and cities as you can. Use the names of football teams or the location of television shows you have seen to help you create your list. Now try to identify a traditional industry that each settlement was originally associated with.

5 Draw a diagram to show the movements of people that were taking place in the UK during the 1970s and 1980s. Your diagram could include the inner areas and suburbs of cities, rural land in the commuter belt and remote rural regions further away. Who was moving in and out of all these different places? Where were they going? What changes did their movement bring?

FIELDWORK FOCUS

The topic of place characteristics, dynamics and connections can be used to frame any number of interesting and potentially unique A-level independent investigations. There is plenty of scope to collect both qualitative and quantitative primary data. A wide variety of resources can be drawn on as forms of secondary data, including literature, music and art. Some possible suggestions are as follows.

A *Profiling different neighbourhoods in order to investigate how and why the demographic characteristics of places vary.* Governments collect a range of data about places that are available in numerical form and can be subjected to statistical tests. These data could be used to compare levels of ethnic diversity in different neighbourhoods, for instance. The secondary data could be complemented with primary interview data: a questionnaire might focus on the reasons why certain social groups prefer to live in some neighbourhoods but not others.

B *Interviewing multi-generational families in order to investigate place connections.* A series of interviews can be arranged with the children, parents and grandparents within several family groups belonging to a diaspora population (see page 15) population, such as British citizens of Indian descent. Data can be collected by asking individuals a range of questions, including how often they have visited India and how often they correspond with family members there using social or traditional media. With more recently established diaspora populations, such as Polish citizens living in the UK, place connectivity could be investigated by asking people about the importance and role of remittances within their family networks.

Further reading

Castells, M. (1996) *The Rise of the Network Society*. Oxford: Blackwell.

Crang, P. (2014) Local-Global. In: P. Cloke, P. Crang and M. Goodwin, eds. *Introducing Human Geographies*. Abingdon: Routledge.

Creswell, T. (2014) Place: *A short introduction*. Oxford: Wiley Blackwell.

Hubbard, P. (2010) *Key Thinkers on Space and Place*. London: Sage.

Jones, C. (2013) How economic change alters nature of work in UK over 170 years. Financial Times [online]. Available at: www.ft.com/content/ebf7ea62-cdfe-11e2-a13e-00144feab7de [Accessed 23 March 2018].

Macfarlane, R. (2015) *Landmarks*. London: Penguin.

Massey, D. (1993) Power geometry and a progressive sense of place. In: J. Bird et al., eds. *Mapping the Futures*. London: Routledge.

Massey, D. (1995) The conceptualisation of place. In: D. Massey and J. Pat, eds. *A Place in the World?* Oxford: Open University/Oxford University Press.

Place meanings, representations and experiences

The experience of belonging to a place can shape our own sense of identity. In turn, people portray places in selective ways through the culture they create. Using a range of qualitative data and place contexts drawn mostly from the UK, this chapter:

- explores the subjective and shared meanings which people attach to places
- investigates how places are represented in different media, and the real-world implications of these representations
- analyses the varied ways in which cities and the countryside have been represented in popular culture
- assesses the tension and conflict that sometimes results from the varied meanings which different groups of people attach to places.

KEY CONCEPTS

Place meaning The significance, value or issues which make a place important for an individual (personal meanings) or group (shared meanings). For example, some places are associated positively with historic events and become a symbol of group identity, whereas others are viewed as 'problem places' for which solutions are needed.

Place representations The diverse ways in which places are portrayed in different kinds of media, ranging from films to personal photographs and diaries. Place representations may include – among other elements – physical landscapes, important landmarks and people.

Contested place A place where tension or conflict has arisen due to the inability of different players to agree how it should be managed, used or represented.

 # Place meanings for individuals and societies

▶ *What kinds of meanings do different people attach to particular places, and why?*

Place attachments and experiences

In contrast with the previous chapter's exploration of objective ('real') place characteristics (such as the site and function of a village or town), the focus turns now to people's more subjective ('imagined') feelings about places. These feelings may derive from:

- our own present or past attachments to places where we have lived or worked
- our own past experiences of visiting particular places
- place representations that *other* people have created (based on *their* own personal experiences and ideas).

Our feelings are what give rise to subjective and personal place meanings. Before reading any further, it is worth taking the time to spend a few minutes reflecting on one or more local places that you know well and the meanings *you* attach to them. Think about: your current home; any places you have lived in previously; holiday destinations; places where you have worked or studied. In your view, what are the defining characteristics of these places? Might other people characterise these same places in contrasting ways based on different experiences of their own? For example, not everyone enjoys their time at school: some students enjoy being part of a community, while others may feel socially isolated or imprisoned within the same place.

Place meanings are, of course, often shared: iconic British landmarks and landscapes, such as Big Ben or the White Cliffs of Dover, are viewed as important by many UK citizens on account of the way they serve as **signifiers** of national identity (see Figure 2.1).

KEY TERM

Signifier Something that represents an important meaning, concept or idea to a group of people who share the same culture..

▼ **Figure 2.1** In First World War propaganda, images of the British countryside were used by the government to arouse patriotic feelings

Identifying positively with places

'To be human is to live in a world that is filled with significant places' (Ted Relph, 1976).

The places that we identify with most positively are those with which we feel a strong sense of connection. Additionally, the aesthetic quality of a place can affect our feelings about it. Vast quantities of landscape photographs are posted daily on social media platforms such as Instagram (see Figure 2.2). Some views have great power to make people reach instinctively for their cameras.

A discussion of the places that people feel they belong to invites critical reflection about the *scale* of our geographical attachments. For example, your own sense of personal identity may be grounded in feelings of belonging to:

- *a home street or neighbourhood* (some people grow up experiencing a fierce sense of rivalry between their own local community and other neighbouring city areas or rural villages; this theme is returned to on page 68)

- *the city*, or part of a city, where your neighbourhood is located (football supporters in large cities with more than one major team may identify with the part of the city which falls within the sphere of influence of the side they follow)
- *the county* that your town, city or village belongs to (for some people, 'Yorkshire' or 'Essex' might be the first answer they give when asked: 'Where do you come from?')
- *your nation* (for many British people, their identity as a Welsh, Scottish, Northern Irish or English person may be of primary importance, perhaps more so than their political status as a passport-holding UK citizen).

> 🗝 **KEY TERM**
>
> **Sphere of influence** The area over which an urban centre or individual business distributes services or draws its customers and visitors from.

There are further ways in which people identify with local places and larger-scale territories. First, increasing numbers of UK citizens belong to a diaspora population (see page 15), meaning that some or all of their ancestors were born, for example, in parts of India, Jamaica, Ireland or Poland. This gives rise to an even more complex sense of personal identity that is rooted in home places spanning two or more different countries. Second, it is possible to feel a strong sense of belonging to a certain physical landscape (rather than the political, social or cultural identity of that place). The classical music composer Edward Elgar often wrote of his feeling of 'belonging' to the Malvern Hills, for instance (see Figure 2.3).

The importance of local places in a globalised world
In the late 1980s and early 1990s, a renewed interest in place meanings and identity politics featured in the writing of many prominent academic geographers, including Doreen Massey, Peter Jackson and Stephen Daniels.

- Their work – which drew in part on theories developed within the disciplines of Cultural Studies and Media Studies – focused on the enduring social significance of local places with unique identities *despite the acceleration of globalisation.*
- This kind of thinking was very different from what many other writers were saying at the time. One popular alternative view was that globalisation had begun to bring about 'the end of geography' on account of the way shrinking world technologies and deeper political integration (such as the European Union) were shaping a world of 'global citizens'. In the new internet age – so the argument went – narrow, nation-based and place-based identities would soon become a thing of the past.
- Not so, argued Massey, Jackson and others. If anything, the feeling of being attached to a local place is needed more than ever in a fast-globalising world. A 'constant correlation' is how the cultural theorist Michel Foucault described this tension between accelerating globalism and increasing localism. Even as people

▲ **Figure 2.2** Instagram and other social media platforms are awash with place images people have created

▲ **Figure 2.3** The Malvern Hills are said to have provided the inspiration for Edward Elgar's music

enjoy the benefits of unprecedented free movement of goods, migrants and ideas, so too may they reject those same things on the basis that global flows have brought too much change to the local places they feel most attached to.

A brief look around the world today reveals individuals and societies in many different local and national contexts apparently 'retreating' into older place-based or religious identities. Notable world events in 2015, 2016 and 2017 included: the UK referendum result in support of Brexit; Donald Trump's unexpected White House victory (promising US withdrawal from numerous global agreements); Catalonia's declaration of independence from Spain; the narrow result of Scotland's independence referendum; and the continuing destabilisation and fragmentation of large parts of the Middle East and North Africa caused by the rise of Daesh (or so-called ISIS). In each case, issues of territorial and place-based identity are, in the minds of many, of equal or even greater value than the economic gains that global integration is alleged to bring.

Places and cultural identity

'Changing places' is sometimes still categorised as a 'cultural geography' topic (though in reality it has become a more mainstream area of human geography over time). This is because study of the concepts of place meaning and identity ideally requires some underlying understanding of what is meant by culture.

▲ **Figure 2.4** In recent years, collective celebration of the 'Englishness' of St George's flag has become a common spectacle at national sporting events

- Any culture (be it local or national) operates as a shared system of meaning; the famous cultural theorist Raymond Williams described culture even more simply as a 'structure of feeling'. This three-word phrase is insightful: it demonstrates that while our own feelings about anything may be personal, they are by no means always *unique*.
- In fact, *collective* displays of emotion are commonplace when you think about it: from funerals and football matches to weddings and Wimbledon, gatherings of people visibly express *shared* feelings (see Figure 2.4).

The culture(s) we belong to influence our lifestyles and create boundaries for behaviour. These may relate to the way we dress, our manners and customs, including which, if any, animals we eat (or do not eat). Some of these rules are religious (or were rooted in religion in the past and persist today as inherited social norms). A society's architecture is part of its culture too: this is why some buildings and landscapes become highly valued and gain protection (see pages 32–35). Culture is also reinforced through a society's music, art and theatre. At the national scale, 'flows' of British culture – including the works of Shakespeare, Beatrix Potter (see page 55), JK Rowling and The Beatles – are one reason why the UK remains a highly globally-connected state.

Like any 'system', culture can be mapped diagrammatically to show the different traits or components, such as language and music, that allow it to function. Moreover, these traits are dynamic. At the same time as a culture is being passed on from one generation to another, it is often evolving too, especially if there is a high rate of technological change occurring or frequent contact is made with people from other places. Connectivity between different places and their cultures (at varying scales) is an essential cause of cultural change.

Reflect briefly on how your own 'lived' culture in the place where you live differs from that of your grandparents. This could be in in terms of music you enjoy, the clothes you wear, what you eat or the way you speak (Table 2.1 shows one view of how English culture has changed over time). Important reasons why your own culture may differ from that of your grandparents' generation could include:

- the growing cultural influence of US, Japanese or Indian television shows, films, food and music
- investment, advertising and market-building by foreign transnational corporations operating in the entertainment and food sectors
- cultural transfers taking place online using social media platforms (Facebook and YouTube)
- global migration patterns and the contributions to UK cultural life made by successive waves of Asian, Caribbean and, more recently, eastern European arrivals (there may be very interesting contrasts to draw between your own culture and that of your grandparents if they migrated to the UK from elsewhere in the world).

Figure 2.5 shows the ways in which place identity and cultural identity overlap. In particular, it is interesting to explore the role that the physical landscape can play in shaping both place identity *and* cultural identity. Important elements of local culture are informed by the physical geography of places, as we have already seen in Chapter 1 (see Figure 1.5).

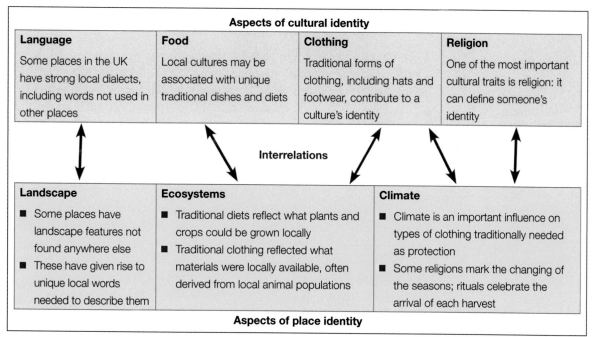

Aspects of cultural identity

Language	Food	Clothing	Religion
Some places in the UK have strong local dialects, including words not used in other places	Local cultures may be associated with unique traditional dishes and diets	Traditional forms of clothing, including hats and footwear, contribute to a culture's identity	One of the most important cultural traits is religion: it can define someone's identity

Interrelations

Landscape	Ecosystems	Climate
▪ Some places have landscape features not found anywhere else ▪ These have given rise to unique local words needed to describe them	▪ Traditional diets reflect what plants and crops could be grown locally ▪ Traditional clothing reflected what materials were locally available, often derived from local animal populations	▪ Climate is an important influence on types of clothing traditionally needed as protection ▪ Some religions mark the changing of the seasons; rituals celebrate the arrival of each harvest

Aspects of place identity

▲ **Figure 2.5** The interrelationships between a society's cultural identity and local place identity help explain why places develop important meanings for the people who live there

	Early twentieth century	Twenty-first century
Religious beliefs	Generally widespread, with high levels of Anglican or Catholic church attendance	Largely secular and non-religious, but with increased religious diversity among those who *do* have faith
Food	Locally-sourced seasonal food, rarely using foreign spices, preferring native herbs	Global, varied tastes in food. Strong spices are widely used in cooking
Identity	People have a strong sense of local belonging (either to a town or county). Regional dialects are stronger than today; most are also extremely patriotic and would fight for their country	Many would be less willing to fight for their country, although they are often strong supporters of national or local football teams. Younger people may see themselves as 'global citizens'
Roots of vocabulary	Celtic, Saxon, Scandinavian (Norse), Roman, Greek, French	Additional Indian, Jamaican, American influences (due to migration, TV and YouTube)

▲ **Table 2.1** How aspects of English culture have changed over time

Place, culture and the environment

A further synoptic example (drawing on elements of both human and physical geography) is the way dried peat (soil with a high carbon content) was traditionally burned as an energy source in some rural regions of the UK, including the Western Isles of Scotland. Peat burning gave rise to highly flavoured 'smoked' meat, fish, and whisky in these places. Today, these traditional food and drink products are valued highly as heritage commodities. They play a pivotal role in the way these places are represented and marketed as tourist destinations.

Robert Macfarlane (see also page 13) has discovered hundreds of different local words for landforms and relief features found only in particular places owing to differences in geology. Traditionally, local cultures have invented words needed to help describe and map their surroundings. For example, the word 'zawn' describes 'wave-smashed chasms in cliffs' and is used only by people in coastal Cornwall living in places where the shore is routinely pounded by high-energy waves with a long fetch. 'Words are grained into our landscapes, and landscapes grained into our words,' notes MacFarlane.

Identifying with or against different places and cultures

People identify positively with some places and cultures. But they may also identify negatively *against* others. As later parts of this chapter show, one reason why place meanings *matter* is because they can be a cause of cultural tension or conflict between different groups of people (see page 67). At varying scales – from gang violence between rival neighbourhoods to the prejudices that citizens of different nations sometimes exhibit towards one another – issues may arise when people experience an *overly* strong sense of belonging to one place rather than another. This can lead to them feeling alienated from, and essentially opposed to, other places. Table 2.2 outlines two important theories about the way processes of geographical and cultural identification operate.

Theory	Explanation
Identifying *with* a place	■ A good starting point for study is Ben Anderson's (1983) account of how every nation is an 'imagined community'. This is widely regarded as essential reading in human geography, more so than ever given current tensions in the UK over Brexit and Scottish independence. Anderson argued that cultures are constructed selectively around particular stories, myths and experiences. Sometimes the narrative is also grounded in the idea of indigenous people but this is not a prerequisite for national identity – as evidenced by the fact that so many Americans view their own national identity as being entirely distinct from Native American culture. ■ Anderson's theory can be applied at the local place scale too: many schools maintain a strong sense of identity by creating a historical 'narrative' that fosters a shared sense of community and belonging among the pupils. This narrative often includes: rooms named after retired head teachers; important annual calendar dates like prize-giving days; traditional uniforms; a 'school song'. ■ In recent years, increasing numbers of people have submitted DNA samples for analysis in order to investigate their ancestral association with particular countries and local places. It has been shown that some people whose ancestors come from Cornwall and Devon have a distinct genetic identity apart from the rest of England, for instance. Other people living in the UK have been surprised to find their roots are more global – and less local – than they had previously thought, following DNA analysis. However, Anderson would argue that a person's identity is about much more than biology: it is about the process of *learning to feel* you belong somewhere.
Identifying *against* a place	■ The connection which people may feel exists between themselves and a particular place is sometimes strengthened *through the act of thinking negatively about other people and places*. A rejection of other places and people (for being 'too different') becomes a way to psychologically strengthen the bond they feel with their own home place. ■ Edward Said's (1978) theory of 'othering' argues that people's sense of identity – both as individuals or as members of a society – emerges through the perception that a difference exists between them and other individuals or societies. Identity is therefore a 'relational' concept. Expressed simply, this means that we may become more self-aware of our own identity when we encounter individuals and societies who are different or 'other' from ourselves. The process Said describes may help explain why such strong hatred can develop between rival gangs belonging to different streets in the same urban neighbourhood. It also contributes to our understanding of why terrible genocides have sometimes been carried out in the distant and more recent past.

▲ **Table 2.2** Theories that help explain how people identify *with* and *against* different places

Why meaning matters

You most likely did not study 'meaning' in GCSE Geography. However, you may have spent a lot of time exploring the meaning or message of poems and novels in GCSE English lessons. In Art, teachers discuss with their students how painters try to represent ideas, while History lessons are sometimes focused on source material showing how historical events were portrayed at the time in different media (and possibly in biased, misleading or strategic ways).

▲ **Figure 2.6** Protests staged to save woodland from being cut down

You can draw on prior teaching and learning in all these other subjects as a jumping-on point for thinking about how and why place meanings and representations *matter*. Places often hold great importance for people, just as some books and films do. A memory of a place can be accompanied by strong feelings and emotions. Indeed, people's attachment to a place can be so strong that they will protest and perhaps physically fight to defend it from change, just as they would act to protect people they love (see Figure 2.6).

Meaning constantly informs and shapes our actions and behaviour in relation to how places are managed.

- Management decisions about places are determined not only by hard facts like unemployment or hazard risk data but also by the feelings and beliefs of people too. Some places are deemed to matter more than others, perhaps on account of important or symbolic landscape features. A cost–benefit decision of whether to install costly sea defences may take account of the hard-to-quantify feelings of stakeholders. People may argue the coastline is too important to lose because its scenery is widely admired or a historic monument is at risk. Perhaps a much-loved film was shot there and people everywhere want the area preserved.

- Of course, the valuing process becomes controversial when some groups fail to understand the value others attach to something. Historic England (see page 33) has awarded protected listed-building status to numerous pieces of modern architecture, including London's Trellick Tower (see Figure 2.7). However, many people cannot understand why these buildings need preserving.

▲ **Figure 2.7** Trellick Tower is a listed building, to the consternation of its critics

- In turn, place meaning determines the marketing strategies used by formal (official) agencies such as tourist boards in the work they do. Decisions must be made about what counts as a valid part of city heritage or culture to celebrate. However, perspectives will differ on what is appropriate: London's Jack the Ripper Museum has been criticised for sensationalising violence against women, for instance.

🔑 **KEY TERMS**

Subculture An alternative system of shared cultural values adopted typically by youthful members of the population. Modern subcultures are typically defined by music, fashion and sometimes strong political views. In the UK, subculture theory is associated with the work of, among others, Stuart Hall and Angela McRobbie.

Technoscape A landscape dominated by state-of-the-art technology and often associated with technology industries. Technoscapes of the past may be preserved, for example Battersea Power Station in London.

CONTEMPORARY CASE STUDY: LONDON'S SOHO

Soho is a place in central London with varying meanings for different groups and **subcultures** both today and in the past. It is a contested place where soaring property values (a three-bedroom flat typically cost £5 million in 2017) have fostered economic redevelopment processes but at the cost of an older cultural landscape rich in artistic and musical history.

The northwest corner of Soho was demolished as part of construction for London's Crossrail network. The Astoria music venue was among many buildings that were lost: between 1976 and 2009, this was one of London's most famous venues and an important place, along with Soho's 100 Club, for the 1970s punk rock movement. In the 1990s, Metallica, Nirvana and Radiohead played at the Astoria. For music fans, this part of Soho was a musical heartland where some of the most famous gigs in musical history were experienced. But this meaning is now lost for future generations.

Also under threat is central Soho's Theatreland. The ongoing changes in Soho have led to the closure of many theatre and revue venues such as Madam Jojo's. A campaign supported by actors Benedict Cumberbatch and Stephen Fry wants to see greater protection given to the unique cultural landscape of Theatreland.

Soho's history suggests it was a place where different subcultures thrived. Part of what made it special to many people was the varied mix of buildings, functions and identities (Soho has also long been an important centre for London's gay communities). Increasingly, however, large parts of Soho and its margins are taking on a more homogenous **technoscape** appearance. The two photographs (see Figure 2.9) show just how much Soho is changing, both in terms of the way it is represented in images such as these and the meaning that these contrasting landscapes have for different groups of people.

▲ **Figure 2.9** Images of Soho and its surroundings: (left) the Astoria in the early 2000s; (right) the new Crossrail station, 2017

KEY TERMS

Text Any culturally-produced representation of something which is considered to contain meaning.

Transcript A record of a depth interview with someone which has been typed out as a document.

Discourse A particular set of arguments or practices through which certain beliefs of meanings are produced and repeated over time.

 ## The power of place representations

▶ *Who decides how places are represented in different media and why does it matter?*

Representing places using different kinds of text

We gain further insight into place meanings and experiences by studying place representations. Cultural geographers use the word **text** to refer to a wide variety of different kinds of place representation. The word can refer to a wide range of qualitative data sources that include – among other things – television, films, music and other sounds, diaries, newspapers, art, cartoons and photographs (see Figure 2.10). When depth interviews are carried out as part of primary data collection, the **transcript** (record of the interview) can also be described as a text.

- Older texts are sometimes described as analogue data (for example, a printed book or vinyl record) and are stored physically in libraries.
- More recent texts are typically digital data: these representations of life are created on a computer or phone and uploaded to an online platform.

You might initially be surprised to find yourself studying 'texts' and 'representations' because we typically associate these words with English rather than Geography. You will be far less surprised by the kinds of issues geographers typically investigate using texts though. These include:

- competing perspectives on whether places should be managed or marketed as tourist destinations (and the way different groups of people express their varying views or **discourses** using traditional or online media)
- tension and conflict between different players or stakeholders (for example, in relation to how places have been represented in the media and the real-world implications of these actions for processes, such as migration).

▲ **Figure 2.10** Geographers make use of a wide range of factual and fictional texts in their studies of place meanings. These include archived photographs, diaries and interviews kept in old analogue and newer digital formats

Some case studies and issues you've studied previously in Geography may prove highly relevant for the study of texts and representations. For example, conflicts over coastal defences sometimes relate to their 'ugly' appearance and the way marketing photographs of a shoreline will look if a management plan is given the go-ahead. Functional yet aesthetically displeasing concrete defences may be effective in preventing flooding but harmful to the coastal resort's tourist trade (see Figure 2.11). Place appearances matter!

Encoding and decoding meanings and messages

In order to investigate texts (words, images or sounds) as place representations, cultural geographers share study methods with English, Sociology and Media Studies researchers (in just the same way that physical geographers, biologists and chemists often have similar methodologies). Several key theoretical ideas have helped build cultural geography's 'toolkit' for handling qualitative data.

- In advertising, someone who is promoting a product relies on the audience being able to 'read' images correctly; for instance, we all know that a 'thumbs-up' image means the same thing as 'good' or 'liked' in British and US culture. Similarly, tourist advertisements often include associated and commonly understood imagery that signifies 'place authenticity'. Images of tartan, heather and castles are used to signify Scotland; pictures of bowler hats have a strong association with London even though hardly anyone wears them any more (see Figure 2.12).
- In all forms of cultural production (and not just advertising), meanings are 'encoded' into texts by their authors with the reasonable expectation that most of the audience will successfully 'decode' the intended message (see Figure 2.13).
- The producers of the different texts we study use their own art form's 'language of signs'. For instance, film soundtracks contain musical 'clues': certain sounds or musical expressions alert the audience to the fact that something romantic or scary is about to happen. This kind of communication only works because most people have learned to recognise or 'read' these musical clues.
- What kind of positive musical clues could be included in a short advertising film about a tourist destination? What kind of messages could you send about a place by including particular songs, notes or sounds? Remember that a variety of music and sounds might be needed in order to communicate a range of messages to different target groups of potential tourists. The model shown in Figure 2.13 is a simplification of what often happens. Note that people will often find different meanings in a text from the one meant by the author. There can be meanings in any text beyond those its producer(s) intended.

▲ **Figure 2.11** Conflict over plans to install 'unsightly' coastal defences relate to how a place will be portrayed in tourist advertising materials and other texts

▲ **Figure 2.12** The image of a bowler hat has a strong association with the UK and the city of London in particular

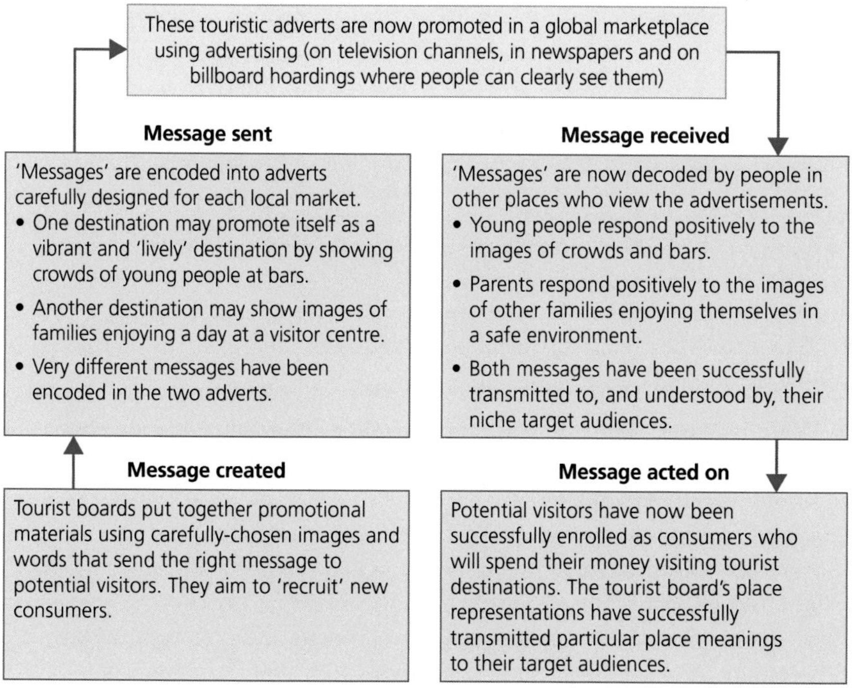

These touristic adverts are now promoted in a global marketplace using advertising (on television channels, in newspapers and on billboard hoardings where people can clearly see them)

Message sent

'Messages' are encoded into adverts carefully designed for each local market.
- One destination may promote itself as a vibrant and 'lively' destination by showing crowds of young people at bars.
- Another destination may show images of families enjoying a day at a visitor centre.
- Very different messages have been encoded in the two adverts.

Message received

'Messages' are now decoded by people in other places who view the advertisements.
- Young people respond positively to the images of crowds and bars.
- Parents respond positively to the images of other families enjoying themselves in a safe environment.
- Both messages have been successfully transmitted to, and understood by, their niche target audiences.

Message created

Tourist boards put together promotional materials using carefully-chosen images and words that send the right message to potential visitors. They aim to 'recruit' new consumers.

Message acted on

Potential visitors have now been successfully enrolled as consumers who will spend their money visiting tourist destinations. The tourist board's place representations have successfully transmitted particular place meanings to their target audiences.

◀ **Figure 2.13** A summary of the main steps of the encoding and decoding process, shown in relation to the production and consumption of tourist advertising literature (note that some readers may of course find an alternative meaning or message to the one intended by the author!)

Different meanings for different people

The business of advertising places is more complicated than simple encoding-decoding modelling sometimes suggests. The old saying 'you can't please all of the people all of the time' is certainly true of place representations. Stop for a minute and think about your own local town centre or another one you are familiar with. Try to imagine an advertising poster for this place using a single photograph that would make it appear highly attractive to everyone, irrespective of age, gender, income or ethnicity. Can it even be done? A vibrant picture of bars and clubs at night might prove attractive to younger people but the same image could alienate or repel an older generation of potential visitors. Equally, a photograph of cafés which cater mainly for older people is unlikely to set the pulses of teenagers racing with excitement at the prospect of a visit.

- Some British cities, such as York and Canterbury, use imagery of their cathedrals in tourist literature. However, modern people vary widely both in terms of their own faith and views about religion in general. How successful would a tourist marketing campaign based entirely on Christian cathedral imagery prove to be?

- Place meanings and representations targeted at one group of people may be contested (challenged) by another group. In the past, many seaside towns relied on poster and postcard images of women in bathing suits as part of their day-to-day advertising (see Figure 2.14). During the 1970s, a feminist backlash against imagery of this kind brought places like Blackpool into disrepute in some people's eyes.

"I must send a postcard to the missus, but I haven't got anything exciting to tell her!"

▲ **Figure 2.14** Some people view the mid-1900s postcards of Donald McGill as humorous representations of life in coastal towns; others find them vulgar and sexist

Carrying out a textual analysis

A geographical analysis of texts – and the discourses (views) of the people who have constructed them – can be carried out sequentially using the following enquiry questions.

1 Who has made this text? Was it produced by an individual (a lone book writer or photographer) or a group of people, such as a tourist board or film production company?
2 Why has this text been produced? Is its main purpose to advertise and market a tourist destination? Could it be that the text is meant as a form of protest? Political images are shared widely on Facebook and other media platforms; many famous books, paintings and photographs have a strong political message, such as Picasso's 1937 painting *Guernica*.

3 Is the text prejudiced or biased in some way and how can we know? Words will tell a story in a particular way; photographs can be framed to leave some people out deliberately or make others look particularly important. Images of historical tourist destination towns such as Stratford-upon-Avon and Oxford are often composed carefully to omit any modern buildings that might spoil the view and ruin the encoded message about historical 'authenticity' that is being communicated to potential tourists. A vital part of any analysis is working out *what has been left out, not just what has been left in.*

4 If the text has a political message, does it reinforce existing inequalities in society or does it try to challenge them? There are many examples of television programmes from the 1970s that reinforced racist views in British society. But there are also plenty of examples of popular songs from the same period, written by multicultural bands like Madness, The Beat and The Specials, that spoke out against racism.

5 Texts dealing with British society in the past and present are sometimes applauded or criticised for the messages they send about diversity. Historical dramas about the First World War have often failed to include any Black or Asian soldiers, despite the fact that very large numbers of them fought for Britain. Actions to tackle this issue of representation are discussed on page 164.

It is important to remember that the skills needed to study texts critically are not developed solely as part of your A-level Geography course. If you cast your mind back to English, Art and other creative subjects in school at previous key stages, what approaches to textual analysis were used?

Carrying out an image analysis

Image analysis can be carried out using the following commonly-used four-stage procedure:

1 *Denotation* – this means identifying and defining the most basic elements of the image.

2 *Connotation* – this means working out how the image (or parts of it) also suggest additional ideas and meanings (a picture of a log fire could be used to represent 'home' for instance).

3 *Mise-en-scène* (literally 'placing on stage') – this involves looking more closely at the subtle ways in which the image has been arranged and framed. Everything in an image contributes to its meaning and message, from people's facial expressions to the clothes they are wearing.

4 *Organisation* – when someone takes a photograph or paints a picture, they decide *what to include and what to avoid.* The composition of the image (for example, some elements are made larger or more in focus than others), the time of day and the angle it is taken from can have an important effect on how an audience responds (see also the examples on page 199).

You can use the above procedure to analyse pictures featured in tourist board websites for possible encoded meanings and messages. Search for images of an area you are familiar with and carry out your own textual analysis.

Formal and informal place representations

The formal and informal parts of any human system are, by definition, quite difficult to pin down. The advent of social media and crowd-sourced data has made the distinction even less clear in recent years. A crude characterisation is that:

- formal place representations are produced by political, social and cultural agencies (including local government, education institutions, tourist boards and national heritage agencies) along with large businesses
- informal place representations are produced by individuals or small groups of people working outside of formal sector institutions. For example, a lone fiction writer or artist can legally say pretty much anything about a place, blurring fact and fiction. Anyone who wants to can write an opinionated online blog about the city or neighbourhood where they live (indeed, a highly biased or prejudiced account may prove attractive to readers who share the writer's views). Informal representations are often creative and do not necessarily try to reproduce reality faithfully.

Formal place representations

Formal or 'institutional' place representations are typically the outcome of a collaboration involving large numbers of people who can be held accountable for what they make. Official tourist brochures, for example, will usually be a product of many people's work and have been checked carefully for factual accuracy. Businesses that publish atlases or city guides may be held accountable for any mistakes they make by their shareholders or owners. Place representations created by formal-sector players are therefore generally characterised as being accurate and objective (with the obvious caveat that tourist literature typically shows the weather of a place selectively, preferring only to use sunny photographs).

Examples of formally-produced materials include the following.

- *Tourist board brochures and websites.* This includes all materials made under the umbrella of the agency VisitBritain, the UK's national tourism agency. This is a non-departmental public body funded by the Department for Digital, Culture, Media and Sport and is therefore answerable directly to the UK Government and electorate.
- *BBC news reports about places.* Under the Royal Charter, the BBC is required to be impartial and unbiased in its reporting. Other major broadcasters are also expected to be objective in their analysis of news events.
- *Published books about places.* This includes your GCSE and A-level Geography textbooks. These books are produced by large successful publishing companies such as Hodder, who have their own reputation to protect; before being published, facts are checked carefully in order to identify and correct inaccuracies or overly biased writing.

Informal place representations

Informal representations are characterised by their 'non-institutional' character and sometimes their means of production too. They are constructed by individuals working outside formal public or private sector institutions, though individual authors and artists will sometimes work in partnership with formal-sector players (publishers and galleries) in order to gain an audience. A *fictional* novel or song about a place might therefore be characterised as an informal representation (even when it is published by a large business). Another example of a partnership could be a tourist board using a local artist's paintings as an advertising image.

KEY TERM

Means of production The physical, non-human inputs into economic systems, such as buildings, machinery and tools. The phrase 'means of production' is associated strongly with the ideas and arguments of Karl Marx.

- The arrival of the internet and social media has allowed informal representations to thrive like never before (whereas, in the past, ordinary people mostly lacked access to the means of production needed to circulate their own informal and independently-made photographs, films or music).
- Informal representations sometimes spread 'virally' online and gain disproportionate influence. Some informally-produced photographs or YouTube videos about places have been seen by millions of people.
- Many best-selling novels or hit songs about places have enjoyed success far beyond the wildest expectations of their creators. Classic songs from the 1960s and 1970s include *Penny Lane* (a song about a street in Liverpool written by The Beatles), *Baker Street* (Gerry Rafferty's song about a London neighbourhood and underground station) and *Ghost Town* by The Specials.

Ghost Town (see Figure 2.15) is a particularly interesting example of an informal place representation. Written in the late 1970s, the song is a critique of how deindustrialisation had adversely affected Coventry's inner city. The song's lyrics conveyed a troubling message about the band's home place. *Ghost Town* became one of the best-selling songs of 1979 and topped the charts in the UK. It sold relatively well in New Zealand and Australia too. This demonstrates how informal representations can be one of the important ways in which local places become connected with distant places (see page 26).

The lyrics of this song are widely available online for you to read. They describe the troubles faced by Coventry at the end of the 1970s. Along with the factories, places of entertainment were closing down. The song describes how fights were becoming commonplace and unemployment was rising – all of which was blamed on the UK Government of the day.

▲ **Figure 2.15** The song *Ghost Town* (1979) by The Specials represented Coventry negatively while also blaming the UK Government for not doing more to help deindustrialised places

Informal texts and subculture theory

In cultural theory, informally-produced texts – particularly popular music – are sometimes associated with particular subcultures. Examples of past musical subcultures in the UK include teddy boys (1950s), mods (1960s), punks (1970s) and new romantics (1980s). Music subculture formation has sometimes had an ethnic or racial dimension: jungle, grime and garage are all music and fashion movements that developed first within the UK's Black community.

Music subcultures are highly relevant to the study of place representations for two important reasons.

- First, music subcultures often have strong home place attachments; for example, Brighton was the focal point for the UK's 1960s mod movement, while Birmingham is widely regarded as the birthplace of British heavy metal music and is now home to a museum celebrating this fact. Britain's music-based subcultures – many of which have gone on to gain global recognition and success over time – have sometimes taken root at first in a solitary urban nightclub. The 100 Club in Soho, London was arguably an origin point for the mid-1970s punk movement; in the 1980s, Manchester's Hacienda Club was widely regarded as the true home place of rave subculture.
- Second, a subculture typically reflects its local environment in the kind of music it makes, either consciously or subconsciously. This is also called making a soundscape (see page 11) of your home place. Thus, the bleak soundscape created by Manchester band Joy Division can be viewed as a reflection of the young musicians' personal experiences of growing up in neighbourhoods torn apart by deindustrialisation in the late 1970s. In the USA, some rap and hip hop music – with aggressive, staccato sounds and sometimes violent lyrics – might be characterised as a soundscape that reflects escalating social and political tensions in deindustrialised neighbourhoods of New York and Los Angeles.

▲ **Figure 2.16** Music subcultures have always thrived in particular home places. The Cavern Club in Liverpool is famous as a venue where The Beatles regularly performed in 1961–2

Harnessing the power of informal representations

Informal place representations – including works of fiction and popular music – can become important cultural resources which formal-sector players, especially tourist boards, later make use of. Many places profit economically from 'local' music, novels or films.

- Music fans continue to visit Penny Lane in Liverpool, half a century after the song about it was written.
- Thousands of people still travel each year to Haworth in rural Yorkshire to experience the real places that inspired the Brontë sisters' novels. Two centuries after they were written, the enduring popularity of books like *Jane Eyre* and *Wuthering Heights* (both first published in 1847) helps support Haworth's economy (also see page 202).
- The UK National Trust maintains children's author Beatrix Potter's house, Hill Top, in the Lake District (see Figure 2.17). Potter wrote some of the *Peter Rabbit* books there in the early 1900s. Under the Trust's management, tens of thousands of visitors each pay £11.50 (2017 prices) to visit Hill Top each year; and there are many beneficial economic impacts for local businesses offering food and accommodation. The site attracts unusually large numbers of Japanese visitors who have travelled halfway around the world to see Beatrix Potter's home. Potter herself could not have predicted the immense popularity and longevity of her little books, least not the long-distance flows of tourists and capital that have made Hill Top such a globally-connected place. There is even a replica of Hill Top in one of Tokyo's parks.

▲ **Figure 2.17** One hundred years after Beatrix Potter's books were written, people still flock to Hill Top to see the real places which her drawings represent. Such is its fame, a facsimile of the building has been built in Japan

Place representations in the social media age

In 2012, the number of registered Facebook users reached 1 billion; the 2 billion threshold was crossed in 2017. Many Facebook and Instagram users post pictures of places regularly. The billion people who use YouTube have created 50,000,000 channels, many of which represent places informally: YouTube users produce content that often provides a glimpse of their surroundings and local neighbourhoods (if only as the background to a homemade film).

Views sometimes differ on whether Facebook and YouTube should be described as formal or informal media providers. This is an important distinction because it affects how they are treated by the law, for instance in relation to libel or hate crimes. One solution is to describe them as 'hybrid' media made up of both formal and informal elements. Facebook is one of the world's largest TNCs and its social media platform is, to some extent, formally managed: the company has established procedures and protocols for dealing with complaints and aims to remove content which is deemed illegal. But at the same time, many Facebook users continue to think of their own news feed as an informal and truly personal view of the world.

Groups of Facebook users sometimes work collaboratively to create their own place representations, a handful of which are shown in Table 2.3. Each is an informal representation which has been co-constructed by a loose network of individual users. Anyone who is a Facebook user can research whether something similar exists for their own home place.

Location	Facebook group name	Number of members/creators (January 2018)
Liverpool	'Crosby Past and Present'	4300 people sharing Crosby neighbourhood photographs
London	'Balham SW12'	2200 people sharing 1990s Balham neighbourhood pictures
Glasgow	'I lived in the old rows in Blairhall'	600 people sharing local neighbourhood memories
Cardiff	'Made in Roath'	4000 people celebrating the rich artistic community of Roath Park, Cardiff
Swansea	'Swansea and its History'	1000 people sharing thoughts and pictures about people and places in Swansea

▲ **Table 2.3** Selected informal place representations created by groups of Facebook users

Place representation on Wikipedia

Among the billions of people who visit Wikipedia regularly, there are around 30 million registered users who contribute or edit its actual content. Wikipedia has become the default source of information for almost anyone online; as a result, its place representations carry enormous global influence. Take time to read the Wikipedia entries for some places you are familiar with. You can carry out a quick, rudimentary content analysis for these place representations by attempting to answer the following questions:

- What kind of geographic, historical, social, economic and cultural information is provided about these places?
- Has any information that you view as particularly important been *omitted* from Wikipedia's representation of your home place or other places? What could be the reason for this omission?
- Can you find out *who* wrote your home place's Wikipedia page and when it was last updated?
- Overall, do you regard the entry for your home place on Wikipedia as a formal or informal place representation? Although Wikipedia operates on a very large scale, it follows different rules and conventions from traditional publishing businesses or government-run providers of information such as the UK's Office for National Statistics.

Some Wikipedia entries lack verification for authenticity and there is uncertainty over their accuracy or who has written them; some contributors remain anonymous.

Place representation on Google Earth

Google Earth is a computer program which has evolved since 2001 into an indispensable app. During this same time, Google has grown to become one of the world's most powerful TNCs. All of the base maps used in Google Earth are copyright-protected satellite images formally produced by NASA and Landsat. However, if you visit your home place on Google Earth you will also see images *informally* uploaded by individual users.

Biased and 'fake' place representations

Although a hyper-connected world brings many benefits, it also enables the rapid and viral spread of information that is either intentionally or unintentionally misleading or provocative, with sometimes serious consequences. Many users of Facebook and Twitter routinely share fake news stories with one another that are completely untrue. Fake news is different from political bias in the media.

- Biased reporting does not set out to tell deliberate lies but instead deliberately omits 'inconvenient' facts or evidence in order to strengthen a partisan (particular) viewpoint.
- Fake news takes things a step further; it consists of *completely inaccurate* or made-up information that has been written and presented (usually online) in a way that makes it appear to be an authentic (and supposedly truthful) mainstream news story. Some fake news reports are politically motivated while others have been written purely for financial gain.

A well-known series of 'fake place' representations were shared widely in 2016.

- Facebook users began sharing reports of how very famous people had experienced flat tyres in small American towns and later praised the friendliness of those places where they had become stranded.
- The articles featured a revolving list of celebrities including Tom Hanks, Adam Sandler, Harrison Ford and Will Ferrell. The places involved included Rochester (New Hampshire), Pflugerville (Texas) and Marion (Ohio).
- All of these stories – which received millions of likes and shares – were generated by two websites whose authors' motivations are still unclear but may have involved generating advertising revenues.

Spotting biased or 'fake' place representations

It is much easier to collect secondary data than it used to be. In the 1990s, most Geography students still had to physically visit libraries when searching for background information to support their independent investigations. Research can be done online now with the click of a mouse. However, the fake news controversy shows how important it is to keep a record of who has authored

 KEY TERM

Hyper-connected A system whose connections have increased to the point where the linkages between system elements (such as people and places) have become incredibly dense.

KEY TERM

Sourcing Identifying and acknowledging the person, organisation or publication from which information has been acquired..

any information you use (this is called sourcing your data). The websites you visit may differ greatly in terms of their credibility and trustworthiness.

It is good practice to always question the validity of information found on the internet. After all, anyone with the right technical skills can build a professional-looking website. Figure 2.18 shows important questions you can ask when carrying out your own research. Two particularly important rules are as follows.

- Don't assume the credibility of a source is shown by how high a story appears in Google search results. Fake news is often very popular and is ranked highly accordingly.
- Be aware of your own biases: we tend to fall for fake news stories more easily when the story corresponds with our own world view.

Fake or genuine?

Authenticity
- What is the name of the news or information provider, and have you heard of them before?
- Can the facts be verified elsewhere? If a Facebook news feed story tells you that the average London house price is now £3.2 million, carry out an online search for 'London house price £3.2 million' and see if a reputable news provider such as the BBC is carrying a similar story (it won't be!).
- Do the facts make sense to you? Common sense tells us that the headline '14 million eastern Europeans are living in the UK' is just plain wrong!

Feedback
- Are contact and author details given, such as an email or business address? Or is the site anonymous?
- Are there links to other reputable sites?

Formality
- Have the designers created an aura of authenticity? Or is the site amateurish?
- Are maps, photographs, fonts used in a polished way?

Personality
- Are personal emotions conveyed or is the feel of the website more professional?
- Can you detect any bias in the way material has been written and presented?

▶ **Figure 2.18** Fake or genuine? Ways of assessing the trustworthiness of online data and news sources

③ Representations of the city and the countryside in popular culture

▶ *In what ways have urban and rural places been represented in present-day and historical media?*

The city and the countryside have been represented artistically in varying ways over time, providing insight into how rural and urban places were experienced by societies of the past. For example, great English literary classics of the 1800s can be crudely divided into:

- urban representations (including the works of Charles Dickens; also Arthur Conan Doyle's *Sherlock Holmes* stories and the early writing of Arnold Bennett)
- rural representations (the novels of Jane Austen, the Brontë sisters and Thomas Hardy).

These representations should not be treated as entirely factual historical records, however. Some Victorian writers portrayed the city and countryside in ways that reflected their own rather narrow experiences of different places and societies. Others wrote in ways which mirrored their political beliefs. Dickens' portrayal of urban life is sometimes rather negative, for instance, echoing his own belief that not enough had been done to help the poor of Victorian London. In contrast, Jane Austen's novels have been criticised for their rather 'rosy' representation of the countryside at a time when large numbers of rural labourers lived in shockingly poor and wretched conditions.

This section now looks in greater depth at the contrasting messages about rural and urban places encoded in different texts and the varied ways in which the city and countryside continue to be represented by today's popular culture.

 KEY TERM

Popular culture The mainstream culture of mass entertainment media, including films, television programmes and popular music.

Urban representations in popular culture

Cities have a split personality if we are to judge by the varied ways in which urban neighbourhoods are represented in modern media.

- On the one hand, cities are often portrayed as sites of progress, modernity and science where people can live prosperous and fulfilling lives ('utopian' places).
- On the other hand, cities are frequently represented as dangerous areas and dysfunctional places where people are threatened by issues such as crime, illness, social exclusion, unemployment and pollution ('dystopian' places).

Utopian urban places

According to the great cultural theorist Raymond Williams, if the rural landscape represents the past then the city can symbolise the future. For some Victorians, cities were state-of-the-art places where engineering marvels like London's Tower Bridge could be admired; their urban architecture continually reminded them of just how far society had come since the start of the Industrial Revolution in the previous century. In contrast, undeveloped rural regions of England at this time still lacked vital infrastructure such as paved roads, streetlights, piped water and sewers.

- The countryside therefore represented a past many Victorians were glad to have left behind.

KEY TERMS

Financescape A statuesque landscape composed of corporate tower blocks and offices such as Canary Wharf in London, the Taipei 101 in Taiwan or the Pudong financial district in Shanghai. Arguably, these landscapes represent not only developmental progress but also the cultural importance of money.

Social housing Rented accommodation provided by local government for people on low incomes, typically blocks of flats in inner-urban areas dating from the 1950–70s.

- It is true that nineteenth-century urban areas had health problems too, but they were also places where modern medicine, health care and plumbing were becoming available for those who could afford these services.

The contemporary city landscape shown in Figure 2.19 epitomises the enduring spirit of what is sometimes called 'heroic urbanism'. Like the Victorians' bridges, mansions and great museums, twenty-first-century skyscrapers and tower blocks make use of designs and materials which are purposely made to impress an audience (such as prospective foreign investors or other countries' governments).

- London's skyline has been transformed again in recent years through the addition of sleek, modern and smart buildings such as the futuristic Shard and the Walkie-Talkie (see page 141).
- Cities compete on the global stage for the prestige of possessing the world's tallest building. This accolade – held since 2010 by Dubai – enhances the reputation of the city that holds the title by representing it as a significant place associated with advancement and progress.
- Futuristic new urban landscapes have been characterised by some theorists as **financescapes** and technoscapes (see pages 47 and 141).

▲ **Figure 2.19** New landscapes in Cardiff (left) and East London (right) embrace futuristic and progressive designs, creating a so-called technoscape which, in some people's eyes, represents development and advancement

Dystopian urban places

Not all people welcome the constant updating and modernisation of city landscapes. An earlier generation of state-of-the-art high-rise buildings in European cities has become widely derided for being ugly and alienating. Brutalism, as page 32 explained, is the collective name for the futuristic concrete architecture that spread across the globe 'like a great fungus between the late 1950s and 1970s', according to architect Edwin Heathcote.

- **Social housing** built by councils to rehouse people living in decaying Victorian slums often adopted brutalist designs. Famous examples include Trellick Tower (see page 46), Red Road Flats in Glasgow and Balfron Tower in east London. Some of the UK's new towns built in the 1950s and 1960s are home to similar structures, as are many university campuses, including some Oxford colleges such as St Edmund Hall and Somerville.

- But many brutalist buildings later proved to be undesirable places to live; rising unemployment disproportionately affected socially-housed populations in inner cities during the 1970s and 1980s, and many buildings exhibited symptoms of social failure such as high levels of crime, family breakdown and drug dealing. Media representations of city life reflected and amplified a growing **moral panic** that urban tower blocks were becoming lawless and 'unliveable' places; science-fiction writers represented brutalist environments using increasingly violent and anarchic images (see Figure 2.20).

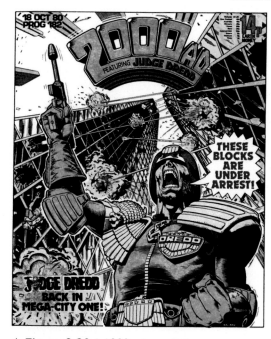

▲ **Figure 2.20** A 1980 science-fiction representation of a dystopian urban place

- Look back at the image of Trellick Tower in Figure 2.7. Critics of brutalist architecture would say it appears a dull, uncaring and possibly hostile place. What's your view?

Derided in the media as 'concrete monstrosities', some blocks like Glasgow's Red Road Flats have now been demolished. Others survived and continue to divide public opinion with regard to their aesthetics. The developer Urban Splash recently renovated Park Hill estate (1961) in Sheffield (page 25) to great applause; even Trellick Tower (originally designed by Hungarian architect Ernő Goldfinger) has been updated inside and is now represented as a fashionable place to live by certain newspapers, including London's *Evening Standard* (a two-bedroom flat typically cost half a million pounds in 2017). Both buildings have received Grade II* listed building status from government agency Historic England.

Urban fearscapes

Negative perceptions and representations of urban places date back much further than brutalism. In successive historical periods, urban places have been understood and represented as **fearscapes**. Ever since British cities first began to grow rapidly in size during the late 1700s and 1800s, newspapers, fiction writers and, more recently, broadcasters have provided their audiences with a steady diet of moral panics, including health scares and fears of violent crime. Although the details have changed over time, fear and panic remain constant.

- Despite being sites of social and technological progress, early Victorian cities were hugely unequal places where scarcity and want were widespread. Low-income groups in overpopulated inner-city neighbourhoods suffered

KEY TERMS

Moral panic An episode of widespread public anxiety and alarm about people's behaviour that has been fuelled by exaggerated media representations of the issue.

Fearscape A place that is experienced or represented in ways that bring anxiety and fear to people.

from outbreaks of cholera and squalid conditions. London newspapers dating from the later 1800s are filled with salacious reports of violent crimes – including the notorious Whitechapel murders – and alcohol abuse (see Figure 1.12, page 22). Is it any surprise that wealthier Londoners often sought to escape to what were then the outer suburbs of Hampstead (see page 18) and Norwood? Another 'axis of fear' in Victorian London ran between the west and east of the city: most well-to-do residents of Kensington and Chelsea were unlikely to venture into the working-class East End.

- By the early 1900s, health and hygiene had improved in urban areas but there were new fears of organised crime for newspapers to focus on. In recent years, the BBC television series *Peaky Blinders*, set in Small Heath, Birmingham, has provided modern audiences with a vivid mythologised representation of this particular moral panic. You may also be familiar with the way famous gangsters such as Al Capone brought fear to the streets of US cities including Chicago during this same period.

- Moral panics about the state of British cities remain a regular staple of news reporting in the twenty-first century. London in particular continues to be portrayed in many sensationalised news reports as a lawless frontier where knife crime and acid attacks are increasingly commonplace.

The study of fearscapes also provides a useful reminder of the subjective way in which people's attitudes towards places vary according to their age, ethnicity, gender and length of residence. Studies of gender in geography have sometimes focused on what has been called 'the geography of women's fear'. This deals specifically with the reluctance of some women to walk home alone at night through particular streets or neighbourhoods on account of poor street lighting or other place features which amplify the sense of personal risk. Can you think of other ways in which personal identity and fearscape issues might be related?

Rural representations in popular culture

Like urban places, the countryside is also represented in contrasting ways by popular culture. On the one hand there is the rural idyll: a romanticised and sentimental 'chocolate box' vision of rural places populated by happy, welcoming and inclusive communities. In contrast to this, the countryside is also at times represented as poverty-stricken or else as a wild and dangerous fearscape characterised by natural hazards, wild animals and unfriendly or even terrorising people and forces (see Figure 2.21).

▲ **Figure 2.21** Contrasting representations of rural places: the rural idyll (left) and a rural fearscape (right)

The rural idyll (positive representations of the countryside)

Pause for a moment and think about positive representations of countryside places and communities in famous novels, paintings or films you know about. Table 2.4 shows how the rural idyll is frequently depicted in British culture and media. At a time when national cultures are changing rapidly, in part because of globalisation, the countryside can also signify a reassuring sense of stability and permanence (see Figure 2.1, page 40). Rural landscapes represent a comforting vision of the past in many people's imagination.

Medium	Example
TV and film	The rural village appears in family-friendly television shows and films as a place where neighbours are good friends who leave their doors unlocked at night. You may have grown up watching *Postman Pat* or *Balamory*, both of which represent rural places in idealised ways. Many adults enjoy comforting film adaptions of Jane Austen's rural novels.
Art	The rural landscape paintings of Constable and Gainsborough portray beautiful, idyllic places.
Literature	As far back as the Ancient Greeks, country living has been praised in European literature. Theocritus and Virgil would later influence Milton and Tennyson. The rural writing of Thomas Hardy remains popular to this day. In a European context, the geographer David Harvey has analysed the enduring importance of the 'Black Forest farmhouse' idyll in German literature.
Advertising	Advertisements for rural holidays in *Country Living* magazine assert there is a positive relationship between moving to a rural place and gaining an improved quality of life. A typical tourist advert might promise readers the chance to 'escape to blissfully tranquil countryside'.
National culture	Some important national myths have a strong association with a rural sense of place, such as the Arthurian legends and the hymn 'Jerusalem'.

▲ **Table 2.4** The countryside is frequently represented as a rural idyll in British culture

The rural fearscape (negative representations of the countryside)

The countryside is alternatively represented frequently as a deprived, dangerous or deranged place – somewhere best escaped *from* rather than somewhere to move to. For many decades, formal representations of UK rural life produced by government and regional development agencies have reported higher-than-average levels of poverty, unemployment and underemployment in some rural 'problem regions' such as the Highlands of Scotland. The Welsh Government has identified rural poverty as a priority area for researchers.

- Retail, health and education services are often lacking in rural places where population numbers fail to meet the threshold required for their provision.
- Lack of access to broadband is another issue that can lead to some remote rural areas being portrayed as excluded and marginalised places.

ANALYSIS AND INTERPRETATION

Study Figure 2.22, which shows push and pull factors that could influence a young person's decision to migrate from the countryside to the city.

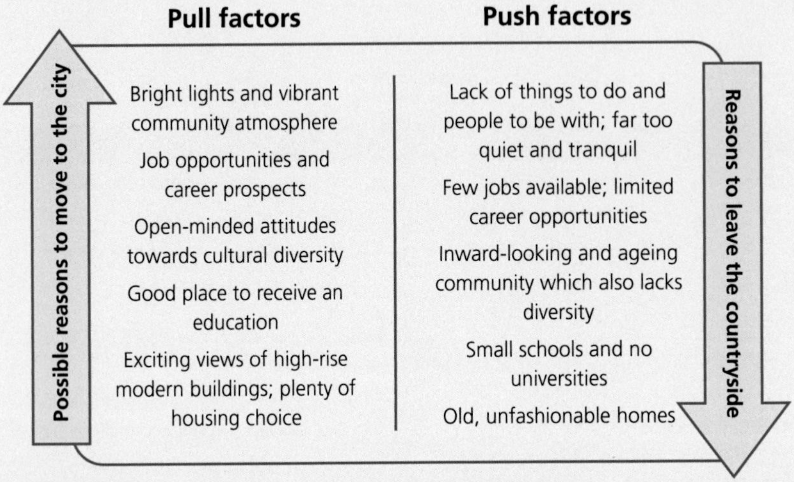

Pull factors   **Push factors**

Possible reasons to move to the city

Bright lights and vibrant community atmosphere

Job opportunities and career prospects

Open-minded attitudes towards cultural diversity

Good place to receive an education

Exciting views of high-rise modern buildings; plenty of housing choice

Lack of things to do and people to be with; far too quiet and tranquil

Few jobs available; limited career opportunities

Inward-looking and ageing community which also lacks diversity

Small schools and no universities

Old, unfashionable homes

Reasons to leave the countryside

▲ **Figure 2.22** The migratory decision-making process for a young person leaving a remote UK rural area

(a) Assess the extent to which the push factors shown in Figure 2.22 are statements of fact or opinion.

GUIDANCE

Five push factors are shown. Both real and perceived qualities of rural life are bundled together here. Can you separate them? Some are statements of truth, while others are statements of opinion or contain mixed elements of fact and opinion. Identify the opinions and explain why they are not statements of fact. Offer a final assessment of the overall importance of opinions and facts for the decision-making process shown here.

(b) Suggest why people of varying ages might have differing views about the pull factors shown in Figure 2.22.

GUIDANCE

Figure 2.22 shows urban place meanings that are bound up with a younger person's identity and experience of life. Consider what different views of urban life may be held by an older parent of young children who has lived in the city for many years. What might this older person think about the perceptions of urban life shown here?

In popular culture, television programmes and films sometimes represent rural places and communities in highly unflattering and even unpleasant ways.

● Alongside the rural idyll, a parallel 'rural tradition' portrays the countryside as a fearscape composed of frightening, lawless, isolated and threatening places. Horror films have long benefited from this trope: many classic scary movies, such as *The Blair Witch Project* (2000), are set in wild places. Crime dramas often employ plots involving psychopaths and murderers in a rural context. Instead of viewing the countryside as a place where you can leave your front door unlocked at night, an audience is left feeling barricades are needed to stop whatever is lurking outside from gaining entry.

- In other portrayals, the countryside is not dangerous but merely boring. From a young person's perspective, this may very well be the case because rural areas lack the entertainment and amenities that urban places provide on account of differences in population density and numbers.

- In some comedy films and television series, including much-loved 1990s show *Father Ted*, rural communities are portrayed as eccentric or even 'backward' when compared with sophisticated city people.

It is important, however, to remain mindful of the diverse rural places that exist in the UK (see page 191). Remoter rural regions are more likely to suffer from the negative perceptions and representations outlined above than are picturesque villages on the rural fringes of Oxford and Cambridge along with other popular commuting areas.

Do city and countryside representations affect migration between rural and urban places?

The push and pull factors that give rise to migration are influenced by perceptions and representations of the source and destination region as much as they are by any fixed reality (a point Everett Lee acknowledged in his famous migration model). As a result, biased representations of rural and urban places really *do* matter. Flattering and unflattering portrayals alike can play a pivotal role in people's migratory or investment decision making (as in the case of a family that relocated from London to rural Cornwall mainly because they 'enjoyed watching *Poldark* on the BBC').

- During the 1970s, major urban centres in the UK lost population. London's population fell from 8 million to 6.5 million between 1961 and 1991; in contrast, many rural regions surrounding major cities (re)gained population. The reasons for this are complex and include deindustrialisation, growing car ownership and the poor state of urban housing in the immediate postwar period. But some movement of urban 'escapees' into rural areas was undoubtedly driven by place perceptions too. Migratory decision-making responsible for the significant counterurbanisation movement occurring during this period may well have been influenced by 'anti-urban' dystopian representations of city life (characterised by Figure 2.18) and 'chocolate box' rural depictions (such as the BBC's long-running *Last of the Summer Wine* show set in the Yorkshire Dales).

 KEY TERM

Counterurbanisation The migration of people from urban to rural areas

- Since the 1990s, major cities have regained population partly due to migration from eastern Europe but also because increasing numbers of UK-born adults are choosing to raise families in central urban areas instead of suburbs or the countryside, new towns or the countryside. Could changing place representations in television, film and other media have played a role in this reversal of fortune for many of the UK's towns and cities?
- In reality, it is impossible to say *exactly* how much influence media place representations have had over time on real flows of people between city neighbourhoods and country places, but they have undoubtedly played an important role.

④ Evaluating the issue

▶ *To what extent can place meanings and representations become a cause of conflict?*

Possible contexts for exploring conflict

Place meanings are the focus of this chapter's final section, including ways in which meanings are translated into representations (such as tourist advertisements, films, music and murals). Do conflicts often arise from the place meanings and representations that different societies create?

The first thing to note when setting the scene for debating this issue is that *a spectrum of tension and conflict exists*, as Figure 2.23 shows.

- Low-level disagreements about place meanings and representations are commonplace. It could be hard to find a marketing campaign for a city, town

or rural area that *nobody* is critical of! Similarly, it is difficult to imagine that every single film or television portrayal of life in a particular place will win the approval of its *entire* audience. This is because people's perceptions of places – and their views about how those places should best be presented to the world – may vary according to their age, ethnicity, gender and length of residence, among other differences. As a result, community tension can always arise over matters of meaning and place representation.

- Real conflict is a very different thing from tension and disagreement, however. Moreover, there are

| Social tension and disagreements over a proposed change to a place | Legal conflicts between different parties, e.g. in relation to new building plans | Isolated acts of vandalism or property damage; violent threats made using social media | Violent street protests or demonstrations; physical assault |

▲ **Figure 2.23** A spectrum of tension and conflict

different ways in which two parties can come into so-called conflict with one another. There are legal conflicts: citizens (and their solicitors) will sometimes make aggressive use of planning laws (see page 33) with the purpose of preserving or removing old buildings that may be widely regarded as essential elements of neighbourhood identity. But there are more physical manifestations of conflict to consider too, ranging from street protests and demonstrations to violent assault and injury. Some examples are provided below of real conflicts arising in particular geographical contexts on account of contested place meanings and representations, often in relation to highly politicised social, ethnic or religious identity issues.

The second important thing to take account of – prior to carefully selecting evidence and examples for analysis – is the way that *tension and conflict can be manifested at different geographical scales*.

- Low-level community tension may arise over a highly localised issue: a particular postcode area might be represented negatively in a regional newspaper article, for instance (perhaps on account of its schools or the quality of housing stock there).
- Far more violent place-based conflicts erupt in local neighbourhoods from time to time too: in most of the UK's major cities, 'turf wars' (open conflict between rival gangs) over local territory fought by rival gangs have sometimes brought loss of life. In Rio and Los Angeles, the annual death toll from turf wars has sometimes reached hundreds or even thousands.
- At the city scale, media representations can lead to dispute, not least because the economic vitality of settlements depends in part on the maintenance of a positive image that helps attract inward investment (an issue which is explored in greater depth in Chapter 4). Negative representations of cities in popular movies and books can lead to tension between citizens and the creators of these materials.

- High-level tension and conflict arises sometimes over even larger-scale issues of place meaning, identity and representation, particularly in relation to the issues of nationalism and sovereignty. Contemporary examples of this include ongoing campaigns for Catalonia's independence from Spain, Scottish independence from the UK and Northern Ireland's reunification with Ireland. These big issues of meaning and identity have been the cause of severe tension and at times violent conflict within particular local neighbourhoods that form part of these larger-scale contested regions.

Evaluating the view that place meanings and representations *are* a cause of real conflict

This chapter previously explored the meaning of the rural idyll (a romantic and old-fashioned view of country life). In the UK, some people strongly associate the rural idyll with the centuries-old tradition of hunting.

- For sections of some rural societies, hunting with hounds is a positive and important symbolic activity which has helped shape their community's identity over time (see Figure 2.24). Yet from the opposing perspective of animal rights, hunting is represented as a cruel and barbaric activity. It is difficult to mediate between these two contrasting positions; for many years, the result was a political stand-off at the legislative level between the pro-hunting lobby (including the Countryside Alliance organisation) and successive UK governments.
- Attempts by Parliament to ban foxhunting led to mass demonstrations in London by hunt supporters in 1997 and 2002. Since 2005, foxhunting has been banned but remains an issue that divides some rural communities. People who used to hunt live foxes continue

Pearson Edexcel | AQA | OCR | WJEC/Eduqas

to gather and ride symbolically though the countryside on horseback accompanied by packs of dogs; sometimes they hunt rabbits.

- At the local level, outbreaks of violence have been recorded both before and since the actual ban on hunting foxes was introduced. In 2015, a saboteur attacked a rider with an iron bar in the Wiltshire village of Everleigh.

▲ **Figure 2.24** Depictions of the rural idyll sometimes incorporate romanticised representations of foxhunting

In urban areas, turf warfare has plagued some local neighbourhoods in major British cities including Manchester (earning it the unwanted nicknames 'Gunchester' and 'Gangchester'), Birmingham and London. Conflict can stem from competing claims over a particular neighbourhood that two rival gangs view as part of their own home place (and whose informal economy they may wish to control, for instance in relation to drug trafficking).

- In Manchester, intense and prolonged gang warfare left at least 40 people dead between 1985 and 2005 as the city's four main gangs – the Cheetham Hillbillies, the Doddington, the Gooch and the Salford Lads – fought to gain control of different parts of the city.
- At Birmingham Crown Court in 2017, 18 members from the notorious Birmingham gangs the Burger Bar Boys and Johnson Crew were made subject to Britain's biggest ever legal

injunction. The gangsters – aged between 19 and 29 – were barred from entering certain parts of Birmingham and were banned from creating and distributing online rap music videos glamorising crime in their local neighbourhoods. The court was determined to stop the gangs from representing their activities and home places in ways that might persuade other young people to join them.

Shankhill in Belfast provides a final example of high-level tension and conflict caused by contrasting and conflicting place meanings and representations. 'The Troubles' was a particularly violent period of conflict between Nationalist and Unionist factions in Northern Ireland that lasted from 1969 to 1998 during which more than 3500 people died.

- In Shankhill and other parts of Belfast, murals on buildings proudly display symbols of identity and events which have special meaning for both groups. In Protestant sections of the city – including Shankhill – paintings of the Union Flag celebrate Northern Ireland's status as part of the United Kingdom (see Figure 2.25), while in Catholic neighbourhoods Irish flags are represented.

▲ **Figure 2.25** Murals depicting conflict painted on buildings serve to represent Belfast neighbourhoods like Shankhill as either Nationalist or Unionist places

- Many murals depict individual heroes who fought or lost their lives in the sectarian civil war. Shankhill's murals include scenes of masked gun-toting 'Loyalist' fighters.

Although the most extreme violence has largely disappeared from Northern Ireland in recent years, tensions remain high between Protestant and Catholic communities. In pursuit of peace and greater cohesion, some communities have removed the most controversial murals and local artists are now painting 'peace walls' instead.

Evaluating the view that place meanings and representations are *not* always a cause of real conflict

In the examples above, issues of geographic and community identity have given rise to real conflict. More usually, however, 'tension' is a better way to describe the feelings generated by contested place meanings and representations.

- Earlier in this chapter, we encountered the 1981 song *Ghost Town*, which had represented Coventry negatively (see page 53). Although it was a huge hit, many people in Coventry were unhappy with the record's success. In an interview, former manager of the band Pete Waterman remembered: '*Ghost Town* created a lot of resentment in Coventry at the time. There were outraged letters to the local paper saying Coventry isn't like that – which of course it was.'
- There are many further examples of informal place representations which are unwelcome in the eyes of local people. Many viewers enjoy watching the ITV drama *Happy Valley*. But some people in Yorkshire's Hebden Bridge, where the series is set, have found the show's focus on crime and drugs distressing because it shows their community in a poor light. In Cornwall, where tourism has received a significant boost from the filming of the BBC series *Poldark*, some residents fear that real-life technology companies may be deterred from investing in a region they now perceive to be quaint and old-fashioned on account of the show's portrayal.

In these and other cases, however, it is important to distinguish between the resentment that media place representations have created for some people and the far more violent forms of conflict discussed previously.

Can tension and conflict be quantified?

Tension and resentment are, of course, difficult things to measure and quantify. Qualitative evidence for the unpopularity of a place marketing campaign might be found in the letters page of a local newspaper or a social media website. But these kinds of data typically consist of isolated letters and comments that could be unrepresentative of the majority of people. How can we hope to make an accurate overall judgement about the level of community tension or conflict caused by issues of place meaning and representation? One attempt was made recently by a team of researchers at Liverpool University.

- They wanted to know if people in Liverpool genuinely approve of how the city centre is being represented as a 'Capital of Culture' and World Heritage Site (WHS) on account of its historic seafront buildings (see page 124 and 144).
- The researchers wondered if the city's WHS status genuinely increases people's sense of pride in Liverpool or whether some locals might dislike this apparent 'obsession' with heritage and the past.
- The results of a survey of 216 citizens are shown in Figure 2.26. Not everyone agreed that WHS status was particularly important; a notably high percentage (41 per cent) disagreed with the proposition. However, although not everyone agreed that WHS status was important (as one of the quotations in Figure 2.26 shows), the researchers found little evidence of people being actively opposed to it.

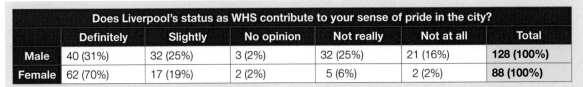

Does Liverpool's status as WHS contribute to your sense of pride in the city?						
	Definitely	Slightly	No opinion	Not really	Not at all	Total
Male	40 (31%)	32 (25%)	3 (2%)	32 (25%)	21 (16%)	128 (100%)
Female	62 (70%)	17 (19%)	2 (2%)	5 (6%)	2 (2%)	88 (100%)

> The World Heritage thing, like, I don't even know when we got it and they kept banging on about it. But if someone took away that certificate, people aren't going to stop coming here 'cause Liverpool's got a good reputation as being somewhere for a good night out.

Respondent 1, city centre

> I think people like the Heritage Society, World Heritage Group, UNESCO and that, can only be good, because without your heritage, you're nothing.

Respondent 2, city centre

▲ **Figure 2.26** Quantitative and qualitative results of a Liverpool University survey showing city centre residents' views about the way Liverpool is now represented globally as a World Heritage Site

It therefore seems reasonable to conclude that recent efforts to represent and promote Liverpool's city centre as a cultural destination have helped to strengthen local community cohesion overall. As we will discover in Chapter 4, the same appears to be true for Hull, the UK's 2017 'City of Culture' (see page 126). In both cities, place meanings and representations which celebrate history and culture are broadly approved of by local communities and do not create tension and conflict. Indeed, actions to regenerate and promote places may actively foster community cohesion in a place by bringing people together and bridging social differences.

Arriving at an evidenced conclusion

Do place meanings *always* become a source of actual conflict? The evidence reviewed above suggests not. Moreover, the answer to that question very much depends upon how you define 'conflict'.

Tensions and disagreements over the use of space are commonplace and are inevitably bound up in the feelings, experiences and perspectives of different stakeholders. Because communities are themselves diverse – if not on account of people's ethnicity and religion then because of their different ages, gender or length of residence – places inevitably have diverse meanings and uses that may not always be very compatible with one another, especially if emotive issues like religion are involved.

Disagreements, tension and conflict are not inevitable, however. Some place meanings and representations may be broadly approved of by all the people living in a locality (although this argument raises the issue of whether we can ever really know this to be true given how difficult it could be to collect supporting data and evidence).

Finally, from a geographer's perspective it is important to always remember that local-scale place tensions and conflicts are often part of a much bigger picture. This is because local places belong to cities or regions that have issues of their own. The meanings and representations that have divided neighbourhoods in Belfast are embedded in the larger-scale contested issue of Northern Ireland's ongoing status as part of the UK.

🔑 **KEY TERMS**

Nationalism The belief held by people belonging to a particular nation that their own interests are more important than those of people from other nations.

Sovereignty The ability of a place and its people to self-govern without any outside interference.

Chapter summary

✔ Places have specific meanings for different individuals and societies. The feelings that we develop about places may derive from actual past experiences of living and working there, or because of how these places have been represented in different media such as books and films.

✔ Everyday place meanings are bound up with different people's personal identities and cultures. Attitudes towards places vary according to our age, ethnicity, gender and length of residence; and also the cultural attitudes we possess that have been shaped by our nationality, religion, values and, increasingly, use of the internet and social media.

✔ Places are represented in a range of different media including films, books, photographs and online creative commons. Formal place representations are created by governments, agencies and large businesses, whereas informal representations can be created by anyone with

a camera or sketchpad. Informal representations can become surprisingly powerful, especially when they are adopted and used by formal sector players.

✔ Representations of places can be biased, inaccurate or fake, and great care must be taken when using and decoding information sources.

✔ Both urban and rural places are represented in contrasting ways using different media to create both positive and negative meanings that can, in turn, influence migration and other social behaviours. Attitudes may vary towards the city and countryside according to people's age or ethnicity.

✔ Place meanings and representations matter because they can become a cause of social tension between different individuals and societies. Issues ranging from religious identity to hunting shape place meanings in ways that have led to actual conflict.

Refresher questions

1 What is meant by the following geographical terms? Place meaning; signifier; culture; text.

2 Using examples, explain how people identify with places at varying scales (use your own experiences or those of other people).

3 Explain why some people's attachment to local places is growing stronger rather than weaker despite the acceleration of globalisation.

4 Using examples, outline how a community's culture is composed of different cultural traits (choose any geographic scale of community, from your school's culture to British culture).

5 Explain the difference between formal and informal place representations.

6 Using examples, explain why some place representations are biased or fake and suggest steps researchers can take to recognise this.

7 What is meant by the following terms? Utopian and dystopian representations; financescape and technoscape; the rural idyll.

8 Using examples, outline reasons why tension and conflict have arisen in some locations on account of contested place meanings and representations.

Discussion activities

1 In pairs, discuss your own attachments to different places that you know personally having lived or visited there. What positive and negative meanings do you associate with these places? How might these feelings influence your decision-making in the future about where you want to live or work?

2 Draw a mind map showing different categories of text (music, films and so on) and add examples you are familiar with (books you have read for example). Add annotations to show what impression you gained about places which featured in these texts. Did the setting for a particular film look attractive and make you want to visit, for example?

3 In groups, discuss how specific places have been represented informally in songs and music. Do these representations matter? What might the real-world consequences be of a particularly positive or negative place representation which features in a song?

4 Discuss the reasons why different individuals and groups of people might find varying meanings in a particular place representation, such as a photograph of the countryside.

5 In pairs, create an assessment of the real-world risks that might be created by fake place representations. What are the possible impacts of completely inaccurate or highly biased adverts and Facebook posts relating to particular cities, societies or neighbourhoods?

6 In groups, discuss your own attitudes towards the city and countryside. Would you describe your own personal views about urban places as 'utopian' or 'dystopian'? Do you view the countryside as an 'idyll' or are your feelings more negative?

FIELDWORK FOCUS

The topic of place meanings and representations provides interesting opportunities for an A-level independent investigation. There is no shortage of secondary data that can be used: think of all the possible novels, photographs, films, websites and tourist board brochures (to name but a few possible sources) that can be analysed and decoded. However, anyone wanting to carry out this kind of work will need to think critically about (i) the best way of sampling place representations and (ii) ways of generating primary data to use in their study.

A *Comparing different place representations in order to investigate contested meanings and signs of conflict.* A series of place representations can be collected, including formal tourist board literature and informal materials (such as photographs and comments about a place which have been posted online). Controversial representations, such as the *Crap Towns* book (see page 12), might also be used. These data can be analysed and decoded to reveal contrasting feelings of the authors. They can also be used to create a visual questionnaire for use with a sample of the general public, thereby allowing primary data to be generated. Respondents can be asked to indicate how strongly they agree or disagree with different representations. The interviews need not be entirely structured: respondents might also be asked to talk at some length about their own feelings towards the place being investigated (the interviews can be recorded using a smartphone app and later transcribed).

B *Using a mixture of questionnaire data and secondary sources to investigate how people experience and understand place boundaries.* A carefully-selected neighbourhood that, preliminary research shows, does not have clearly-defined margins can be chosen for investigation. This is typical in larger towns and cities where local railway stations provide some sense of place identity, but it is unclear where one place ends and another begins (for example, one view of London is that it is composed of hundreds of neighbourhoods built around different underground stations, such as Wimbledon, Clapham Common and Balham; but boundaries between these places can be subjective). The place where you live may be widely regarded as being composed of different neighbourhoods; do views differ on where the boundaries are? Part of the primary research could involve asking interview respondents to annotate a local map to show where they feel the neighbourhood boundaries should be drawn. Another approach might be to ask respondents in a city centre to annotate a map showing where they feel the 'town centre' or 'shopping centre' edges are.

C *Interviewing selected people in order to investigate a particular issue that links people's personal identity with their feelings about local neighbourhoods.* The 'geography of fear' issues outlined on pages 61–63 could serve as the theoretical basis for an investigation. You could interview members of your own school class about local 'no-go areas' where they might feel nervous to walk alone (it might be a good idea to stratify your sample to include equal numbers of male and female respondents); a more positive approach could be to map safe or 'happy' areas instead. Whatever the focus, there are plenty of opportunities to develop a challenging and rewarding data collection programme that invites participants to annotate local maps for you to collate and analyse. You might consider using digital mapping and data analysis; the online 'mappiness' resource may provide inspiration: www.mappiness.org.uk.

Further reading

Anderson, B. (1983) *Imagined Communities: Reflections on the Origin and Spread of Nationalism.* London: Verso.

Creswell T. (2014) Place. In: P. Cloke, P. Crang and M. Goodwin, eds. *Introducing Human Geographies.* Abingdon: Routledge.

Crang, M. (1998) *Cultural Geography.* London: Routledge.

Crang, M. (2014) Representation-reality. In: P. Cloke, P. Crang and M. Goodwin, eds. *Introducing Human Geographies.* Abingdon: Routledge.

Daniels, S. (1993) *Fields of Vision: Landscape and National Identity in England and the United States.* Cambridge: Polity Press.

Hann, M. (2017) The fight to save London's live music scene. *FT Magazine,* 11.

Morley, D. (1992) *Television, Audiences and Cultural Studies.* London: Routledge.

Relph, E. (1976) *Place and Placelessness.* London: Sage Publications.

Rose, G. (2006) *Visual Methodologies: An Introduction to the Interpretation of Visual Materials.* London: Sage.

Said, E.O. (1978) *Orientalism.* New York: Random House.

Schama, S. (1995) *Landscape and Memory.* London: HarperCollins.

Tuan, Y. (1977) *Space and Place.* Minneapolis: University of Minnesota Press.

Valentine, G. (1989) The geography of women's fear. *Area,* 21(4), 385–390.

CHAPTER 3

Place changes, challenges and inequalities

In time, everywhere changes. The geographical processes and connections that originally helped a place thrive may eventually cease to operate. In turn, new external forces constantly reshape places, societies and patterns of inequality. Using a range of examples of industries, locations and communities, this chapter:

- explains how global shift, deindustrialisation and a cycle of deprivation affected British towns and cities in the second half of the twentieth century
- explores new twenty-first-century economic, geopolitical and technological challenges for places
- analyses the causes and consequences of demographic and cultural changes affecting different places
- assesses the present-day severity of spatial inequalities in the UK at varying scales.

KEY CONCEPTS

Global shift The international relocation of different types of industrial activity, especially manufacturing industries. Since the 1960s, many industries have all but vanished from Europe and North America. Instead, they thrive in Asia, South America and, increasingly, Africa.

Inequality The social and economic (income and wealth) disparities that exist both between and within different places and their societies. Inequalities at national, regional (city) and local scales often give rise to migration.

Cultural diversity The extent to which a population is culturally heterogeneous (varied) or homogenous (the same). Ethnicity, race, religion, language and other real (or perceived) cultural traits can be used as measures of cultural difference and diversity.

Positive feedback The way that socioeconomic or environmental changes become accelerated by the processes operating in a human or physical system.

① Deindustrialisation and the cycle of deprivation

▶ *How did deindustrialisation affect declining UK towns and cities in the later twentieth century?*

Global shift

The term 'global shift' describes the international relocation of different types of activity. Since the 1960s, a range of manufacturing including heavy industry (such as steel making and shipbuilding) and assembly operations (especially electronics) have all but vanished from Europe and North America, along with some primary industries (particularly coal mining in the UK). Instead, these activities now thrive in Asia, South America and, increasingly, African nations such as Ethiopia.

- Global shift is a complex process which includes both **offshoring** and **outsourcing** strategies along with new business start-ups in emerging economies.
- The root cause is the pursuit of profit in a capitalist world economy. Global shift in the postwar decades was the inevitable outcome of cheap labour and land costs in the so-called 'global south' where economic growth had begun to deliver improved infrastructure, rising standards of education and political development strategies such as **export processing zones**.

Although global shift has brought some positive economic changes to many parts of the world, its effects for developed nations have not always been welcomed. The economic, social and environmental winds of change that swept through the UK's industrial heartlands in the 1970s and 1980s became known as deindustrialisation. This broad umbrella term has various meanings. In its broadest sense it describes a sustained decline in the importance of industrial (especially manufacturing) activity for a place. More specifically it may refer to a fall in:

- the relative or absolute importance of manufacturing activity as measured by its contribution to gross domestic product (GDP)
- the relative or absolute percentage (or number) of a place's population who work in traditional industries (manufacturing and primary industries such as coal production).

It is important to remember that the term 'deindustrialisation' is sometimes used by different people to mean different things. As we shall see, some places in the UK have maintained a high manufacturing output over time yet rely increasingly on automated assembly and robots rather than a human workforce. This can lead to a situation where manufacturing output data are healthy but local communities suffer high unemployment.

Figure 3.1 shows a global timeline of the post-1945 period which includes key economic, political and technological influences or 'shocks' whose legacies have significantly reshaped the UK's economic and social geography. Global shift is portrayed in this timeline as an ongoing process which begins with the drift of heavy industry to Asia in the early 1960s, accelerates on account of a succession of economic crises during the 1970s and enters a new chapter with the full emergence of India and China as major global players around the end of the millennium. It is important to recognise that throughout this timeline, economic forces and processes have been *allowed* to operate by **neoliberal** governments and organisations acting at both national and international scales.

🔑 KEY TERMS

Offshoring TNCs move parts of their own production process (factories or offices) to other countries to reduce labour or other costs.

Outsourcing TNCs contract another company to produce the goods and services they need rather than do it themselves. This can result in the growth of complex supply chains.

Export processing zone An industrial area, often near a coastline, where favourable conditions are created to attract foreign TNCs. These conditions include low tax rates and exemption from tariffs and export duties.

Neoliberal A philosophy for managing economies and societies which takes the view that government interference should be kept to a minimum and that problems are best left for market forces to solve.

1944–45 Establishment of the World Bank, International Monetary Fund (IMF) and origin of the World Trade Organization (WTO). The postwar Bretton Woods conference provides the blueprint for a free-market non-protectionist world economy where aid, loans and other assistance become available for countries prepared to follow a set of global financial guidelines written by powerful nations.
1960s Heavy industry in the developed economies of Europe and America is increasingly threatened by rising production in southeast Asia, including Japan and the emerging Asian Tiger economies, notably South Korea. Unionised labour costs push up the price of production for Western shipbuilding, electronics and textiles. Most older economies enter a period of failing profitability for industry.
1970s The first OPEC Oil Crisis of 1973 pushes Western industry over the edge. Rising fuel costs trigger a 'crisis of capitalism' for Europe and America, whose firms begin to step up outsourcing and offshoring of their production to low labour-cost nations. Meanwhile, soaring petrodollar profits for Middle East OPEC nations signal that the United Arab Emirates and Saudi Arabia are on their way to becoming new global hubs. In 1978, China begins economic reforms and opens up its economy.
1980s Financial deregulation in major economies like the United Kingdom and the United States brings a fresh wave of globalisation, this time involving financial services, share dealing and portfolio investment (by 2008, financial markets would have an inflated value more than twice the size of actual world GDP!). The collapse of the Soviet Union in 1989 significantly alters the global geopolitical map, leaving the United States as the only 'superpower'.
1990s Landmark decisions by India (1991) and China (1978) to open up their economies brings further change to the global political map. Established powers strengthen their regional trading alliances, including the European Union (EU, 1993) and North American Free Trade Association (NAFTA, 1994). The late 1990s Asian financial crisis is an early warning of the risks brought by loosely-regulated free-market global capitalism.
2000s Major flaws in the globalised banking sector emerge during the 2008–09 global financial crisis (GFC). Unsecured loans totalling trillions of dollars undermine leading banks. This brings a negative multiplier effect that causes a fall in the value of global gross domestic product (GDP). In an interconnected world, growth slows for the first time in two decades for China and India, the two great outsourcing nations and new emerging superpowers.
2010s Growth remains slow following the GFC, with many countries slipping in and out of recession, including Russia and Brazil. Problems in Greece and Portugal escalate into a Eurozone crisis. Despite slower growth, China overtakes the United States to become the largest economy by purchasing power parity (PPP). In many countries, popular opposition to migration and free trade is on the rise and the United Kingdom votes to leave the EU. However, global internet and social networking use reaches new record levels. Experts struggle to tell whether globalisation is increasing, pausing or retreating.

TIME

▲ **Figure 3.1** A post-1945 timeline for globalisation and global shift of industry

Structural economic changes in the UK

Figure 3.2 shows the structural economic transformation of the UK as a whole since the 1950s. Heavy industries (including iron, steel and shipbuilding) and traditional industries (textiles, chemicals, engineering)

were hit hardest by global shift and world events (see Figure 3.3). The UK's major cities – most of which grew originally in size and status precisely because of their successful traditional industries – haemorrhaged manual work (Table 3.1). Over time, tertiary and quaternary sectors of industry have become far more significant in terms of their contribution to both employment and the UK's national income. However, this broad structural shift – sometimes referred to as tertiarisation – has played out unevenly from place to place.

- Local resilience to changing global economic conditions has been highly variable. Overall, the UK today has a secondary sector which still employs 2 million people and contributed 20 per cent of GDP in 2017 (if construction industries are included): clearly, some places have retained thriving manufacturing industries which employ large numbers of people. Later in this chapter you will read about some of these businesses and their locations (see also Chapter 4, page 135).
- Processes of change were more rapid for some regions and places than others. For example, between 1952 and 1979 manufacturing employment in the northwest of England fell by 25 per cent. This was a far steeper decline than in other regions and reflected the northwest's unfavourable mix of industries (in the context of global shift), which included textiles, chemicals and light engineering.
- In the 1980s, manufacturing decline accelerated in northern regions, while southern England proved more resilient against further losses (see Figure 3.4).

City	How manufacturing declined
Liverpool	By the 1970s, Liverpool's days as the largest port in the UK were numbered. Changes in global trade made Bristol and other ports in the south more popular and unemployment rose by one-third during the 1970s. Two thousand businesses closed between 1978 and 1982, including textiles, engineering and electrical firms. In total, 200,000 jobs disappeared as Liverpool deindustrialised.
Manchester	In 1959, manufacturing still employed over half of the Greater Manchester workforce; today, it accounts for fewer than one in five jobs. Deindustrialisation hit the city hard: between 1971 and 1981, Manchester lost almost 50,000 full-time jobs – particularly in textiles, heavy engineering and chemicals – and almost one-fifth of its population.
Cardiff	During the nineteenth century, Cardiff was the world's second-biggest coal-exporting port. Light industries were located near the docks and Cardiff also became a regional hub for finance and insurance. However, the decline of the Welsh coal and steel industries meant that the port began to suffer during the 1950s and 1960s.

▲ **Table 3.1** The loss of traditional manufacturing industries and employment caused by global shift brought great hardships to large UK cities. No major UK settlement remained immune to the effects

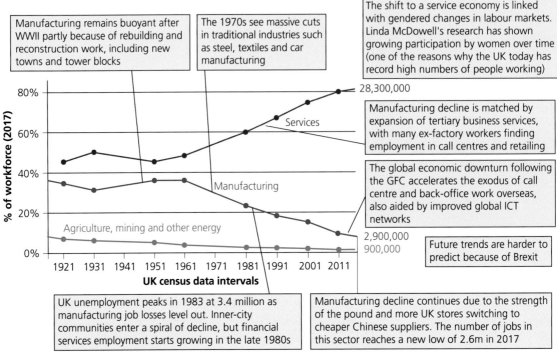

The shift to a service economy is linked with gendered changes in labour markets. Linda McDowell's research has shown growing participation by women over time (one of the reasons why the UK today has record high numbers of people working)

Manufacturing remains buoyant after WWII partly because of rebuilding and reconstruction work, including new towns and tower blocks

The 1970s see massive cuts in traditional industries such as steel, textiles and car manufacturing

Manufacturing decline is matched by expansion of tertiary business services, with many ex-factory workers finding employment in call centres and retailing

The global economic downturn following the GFC accelerates the exodus of call centre and back-office work overseas, also aided by improved global ICT networks

Future trends are harder to predict because of Brexit

28,300,000

2,900,000
900,000

% of workforce (2017) — Services — Manufacturing — Agriculture, mining and other energy

80% — 60% — 40% — 20% — 0%

1921 1931 1941 1951 1961 1971 1981 1991 2001 2011

UK census data intervals

UK unemployment peaks in 1983 at 3.4 million as manufacturing job losses level out. Inner-city communities enter a spiral of decline, but financial services employment starts growing in the late 1980s

Manufacturing decline continues due to the strength of the pound and more UK stores switching to cheaper Chinese suppliers. The number of jobs in this sector reaches a new low of 2.6m in 2017

▲ **Figure 3.2** The changing employment structure of the UK, 1921–present (ONS data)

▲ **Figure 3.3** Industries which thrived in the early twentieth century often found conditions harder in the postwar decades

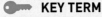 **KEY TERM**

Rationalisation Ways of making an industry more efficient, for instance by shedding labour and trying to increase the productivity of the remaining workers, or closing a company's less profitable operations.

The manufacturing trends shown in Figure 3.2 demonstrate how manufacturing employment has continued to fall steadily in recent decades due to a combination of automation, factory closures and strategic relocation abroad.

● The patriotic UK entrepreneur James Dyson resisted moving his own firm's manufacturing wing to Asia for many years. In 2003, however, profitability concerns finally prompted the Dyson Corporation to move its shop-floor operations to Malaysia in order to reduce labour costs by two-thirds. The move also made sense because of growing sales of Dyson's vacuum cleaners in emerging Asian markets.

● The enlargement of the European Union in 2004 provided an opportunity for UK manufacturers to move operations to lower-wage eastern European countries. Some companies took advantage of EU membership to acquire, or merge with, rival firms headquartered in other countries, at times resulting in **rationalisation** and the closure of UK branches in favour of expanded production at existing sites in other European territories. For example, in 2006 Nestlé announced the loss of 645 jobs at its factory in York. The company moved the production of *Smarties*® to Germany, *KitKat*® to Bulgaria and *Aero*® to the Czech Republic where wages are much lower than the UK.

ANALYSIS AND INTERPRETATION

Study Figure 3.4, which shows how the level of unemployment in different UK regions deviated from the national average between 1974 and 1991. A positive differential figure shows above-average unemployment in a region and a negative differential figure shows below-average unemployment.

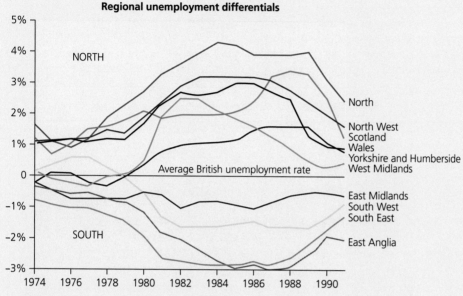

▲ **Figure 3.4** Regional unemployment differentials for UK regions, 1974–91

(a) Using Figure 3.4, calculate the difference between the unemployment rate for East Midlands and West Midlands in 1982.

(b) Compare the different trends for northern and southern regions between 1974 and 1984.

GUIDANCE

The command word 'compare' requires an analytical approach to the data which goes beyond merely listing facts or providing a simple sequence of descriptive statements. Instead, the analysis should be carried out using comparative language throughout (including frequent use of connective words like 'whereas' and 'however'). One way of approaching this task might be to first compare the 'big story' of the north with that of the south over time, prior to providing a more detailed comparison of individual regions. The fact that regional trends diverged markedly after 1978 is worth highlighting. Be careful to describe the data accurately: the graph shows unemployment differentials *compared with the national average* rather than actual rises or falls in the overall unemployment rate. Be careful also not to compare the trends after 1984.

(c) Suggest reasons for these differences.

GUIDANCE

The relatively poor performance of northern regions can be explained with reference to deindustrialisation and the failure or relocation of manufacturing industries. Some mention might also be made of changes in employment for other traditional industries such as coal mining. The fact that southern regions performed relatively less poorly reflects their more diverse economies. Overall, Figure 3.4 shows that regional exposure to the impacts of global shift was very uneven during this time period.

(d) Suggest reasons why regional unemployment differentials begin falling in the late 1980s.

GUIDANCE

This question requires careful thought and application of knowledge. One possible reason could be that measures had been taken to provide new employment in the worst-hit regions by the late 1980s. For example, new inward investment had arrived in the UK, such as Nissan in Sunderland; the UK Government had also taken deliberate steps to incentivise inward investment into problem regions by transnational corporations. Another possible reason might be the increasing importance of service industries to the UK economy during the 1980s. New employment opportunities in retail, tourism and leisure helped to reduce unemployment in many cities that had been hit hard by deindustrialisation in the previous decade.

(e) Assess the strengths and weaknesses of Figure 3.4 as a way of presenting regional unemployment trends.

GUIDANCE

Think about other ways of presenting regional unemployment data. It would have been possible to show the actual level of unemployment in each region instead (as opposed to the level of deviation from the national average). As a result, we do not know whether unemployment was in general rising or falling from year to year: this could be seen as a weakness of Figure 3.4. The strength of this presentation technique is the way that it draws attention to the diverging fortunes of northern and southern regions (the relative resilience of the south is made clearer; so too is the disparate exposure of the north to global shift). You may be able to offer an assessment of other strengths and weaknesses, such as the ease with which the data can be interpreted and understood.

Changes in car-making places

Looking back at Figure 3.2, it is important to remember that some places that are correctly described as having undergone deindustrialisation in terms of people's *employment* may still have a strong manufacturing *output*. This is because of the adoption of mechanised production techniques. Computer-aided design (CAD) and computer-aided manufacturing (CAM) have maintained output at the expense of employment, especially for the UK's car industry.

- In total, 1.7 million cars were built in the UK in 2017, a record number of which were exported worldwide (more than one-in-two went to other European countries): this is the result of investments over recent years in cutting-edge design and technology.
- Although a similar number of cars were built in the UK in 1972, the size of the workforce now is radically smaller. Today, the car industry has a highly skilled and productive workforce numbering just 169,000 in 2016; in 1972, the figure was 500,000.
- Another important change is the way that all large car-manufacturing operations in the UK have passed into foreign ownership as a result of foreign direct investment into the UK by, among others, large German, Chinese and Japanese firms. Numerous iconic British brands have passed into foreign hands in recent years (see Table 3.2).

Perhaps the most well-known car manufacturing hub in the UK today is Sunderland, where the Japanese TNC Nissan has produced cars since 1984. Nissan's inward investment came with strings attached – the UK Government signed an agreement offering land at an incentivised price. Because the northeast of England had already undergone deindustrialisation following the closure of shipyards and coal mines, there was a large workforce for Nissan to draw on. Today, Nissan's Sunderland factory is the largest car plant in the UK, producing about 500,000 vehicles a year and employing 7000 people in 2017. This may still sound a lot but, by comparison, Ford Motors employed more than 40,000 workers at its Dagenham plant in the 1950s.

Brand	Place	Products	Ownership
Aston Martin	Gaydon	Cars	UK
Bentley	Crewe	Cars and engines	Germany
Ford	Bridgend, Dagenham & Southampton	Engines, buses and coaches	USA
Honda	Swindon	Cars and engines	Japan
Jaguar Land Rover	Castle Bromwich, Solihull & Halewood	Cars	India
Lotus	Norwich	Cars	Malaysia
London Taxis (LTI)	Coventry	Cars	UK/China
McLaren	Woking	Cars	UK
MG Motor	Longbridge	Cars	China
Mini	Oxford & Birmingham	Cars and engines	Germany
Morgan	Malvern	Cars	UK
Nissan	Sunderland	Cars and engines	Japan
Rolls-Royce	Goodwood	Cars	Germany
Tata	Coventry	Cars	India
Toyota	Burnaston & Deeside	Cars and engines	Japan
Vauxhall	Ellesmere Port & Luton	Buses and coaches	USA

▲ **Table 3.2** Places in the UK where cars are still manufactured. The global pattern of foreign ownership shows how shifting flows of investment have occurred over time

Table 3.2 shows places in the UK where large numbers of cars are still made. The changes in the car industry outlined above have affected these places where vehicles are made in several important ways.

- In most of the places shown, a lower proportion of local people are directly employed by car industries than in the past because of automation. This can affect place meaning and community identity because fewer people now share the same personal experience of actually building cars together (see Figure 3.5). Traditional industries in the UK rarely function now (in the way they used to) as the 'glue' that binds whole communities together (see also pages 11–12).
- Because of foreign ownership, the identities of places where cars are made have arguably been reshaped by connections with distant places. For example, although Oxford may remain the 'home' of Mini, profit flows are directed ultimately towards Germany (where Mini's owner, BMW, is headquartered).

▲ **Figure 3.5** In the past, before automation, large numbers of young people worked together in local car industries: this shared experience helped foster a strong sense of place and community identity

The UK's decision to leave the European Union has future implications for the places shown in Table 3.2. When German, Japanese, Chinese, Indian, American and Malaysian companies decided to invest in production facilities in the UK, they did so under the assumption that the UK would remain within the EU. At the time of writing, it is still not entirely clear what the terms of any trade deal between the UK and the EU will be in the future, but it is feasible that UK car exports may become more expensive for European customers. Although Nissan and Toyota have pledged to stay in the UK no matter what happens, some companies have sought assurances from the UK Government that they will be supported financially if export costs rise.

Changes in coal-mining places

The UK's major coalfields include South Wales, South Yorkshire, Great Northern and Ayrshire. The coal industry originally shaped an entire way of life in mining towns and villages: the shared experience of working deep underground in dangerous conditions fostered a strong sense of identity and community cohesion.

- Welsh mining communities were renowned for their male voice choirs (see page 12).
- In the early 1900s, the footballing cities of Newcastle and Sunderland owed their pride, prosperity and employment to coal: one of the greatest figures in the history of English football, Bob Paisley, was a miner's son from County Durham.
- The opening chapter of George Orwell's 1937 book *The Road to Wigan Pier* is a valuable qualitative data source for anyone interested in the cultural landscape of coal-mining places.

Today, fewer than 4000 people are employed in the remaining handful of opencast mines still in use (see also page 94). In contrast, there were 1 million miners in 1914. Other than the exhaustion of some coal seams, what explains the loss of this once-great industry?

- Due to the dangerous nature of their work and the large number of people involved, mining communities established strong trade unions. Over time, the National Union of Miners (NUM) successfully gained improved pay and working conditions for miners from private mine owners and then from the government-owned National Coal Board.
- But the rising cost of manual labour in developed countries is, of course, a major reason why global shift and deindustrialisation occurred. In the 1960s and 1970s, successive UK governments attempted to close expensive UK pits with the aim of importing cheaper coal from Asia, South America, Africa and eastern Europe instead.
- Fierce resistance to threatened pit closures from the NUM culminated in a national strike in 1984 among its 180,000 members (around three-quarters of all remaining coal miners at that time). The political standoff between

Margaret Thatcher's Conservative government and the NUM leader Arthur Scargill lasted for a year (see Figure 3.6). Eventually, however, the strike collapsed and many pit closures followed.

- Coal mining has declined even further since the 1980s, most recently because of a 'green' political drive to remove coal altogether from the UK's energy mix to help reduce greenhouse gas emissions.

Many places where coal was mined traditionally continue to struggle economically today. Most mining settlements existed because of coal and for no other reason; it is sometimes difficult to bring new flows of investment to these places and diversify their economies. Cortonwood Colliery near Rotherham in South Yorkshire used to employ a close-knit community of 1000 miners. It has fared better than many other places since the strike of 1984, after which its colliery closed. Proximity to the A1 made Cortonwood a suitable site for a retail park and there is now a supermarket where the miners' communal showers once stood, while a branch of B&Q marks the site of the old mineshaft. Some younger ex-miners have found work in these stores.

Perspectives on changes at Cortonwood and other places differ.

- Ex-miners are often proud of their past – during the two world wars, mining was a 'reserved' occupation, meaning it was essential for national survival. Some regret the end of coal mining because of the strong sense of local identity it created, along with the economic security it gave places.
- Others may think it best that global shift and green politics have consigned coal mining to history in the UK. Mining left men with disabilities; inhalation of coal dust led to high rates of emphysema and lung cancer. One-quarter of working-age men in ex-mining communities receive sickness benefits. Employment changes have also benefited women greatly (given the strongly gendered character of coal-mining work historically). As a result, Millennials born in places like Cortonwood may wonder why the older generation regret the pit closures quite so much.

▲ **Figure 3.6** Economic development processes can bring changes which are bitterly contested by communities: the coal miners' strike of 1984–85 often turned violent and demonstrated how the shared experience of mining had given rise to strong community and place identities

Inner-city challenges and the cycle of deprivation

In some places, the employment problems caused by structural industrial change quickly became amplified to create increasingly severe socioeconomic challenges. The specialised geographical concept of positive feedback helps explain why. You may already be familiar with this idea from studying physical systems in geography. Positive feedback in a system is understood as a magnification of initial effects that ultimately creates energies which spiral out of control. The impacts of deindustrialisation in developed countries were amplified by feedback in several ways during the 1970s and 1980s.

- Each major industrial closure, like the loss of Liverpool's Tate & Lyle factory, had knock-on effects that rippled through supply chains, causing 'upstream' and 'downstream' major industries and smaller suppliers to fail too. Entire clusters of industries disintegrated. In Sheffield, there were 27,000 steel production jobs in 1981. By 1991, only 7000 remained. The linked effects for other sectors of the city's economy were enormous. One-in-four of all the city's jobs was lost, both in producer industries (metal manufacturing and engineering) and the consumer retail sector (whose customers' incomes plummeted following the termination of their employment).

- Feedback effects accelerated the deterioration of the physical environment. The broken windows theory describes the way that small acts of vandalism can spiral out of control into more serious problems like arson. Derelict buildings also attract criminal activities including drug manufacturing. As environmental conditions worsen, any last remaining businesses relocate elsewhere (see Figure 3.7).

- Social problems accelerated through the operation of a cycle of deprivation. As its name suggests, this is a self-sustaining feedback loop (see Figure 3.8). In the worst-affected neighbourhoods of UK and US cities, drug-related crime became the basis of an informal economy in some poor neighbourhoods. The life expectancy of some low-income US urban districts is now 30 years lower than in neighbouring affluent places. Gun crime plays a large part in this statistic. Page 68 explored how UK cities like Manchester also suffered from gang violence.

- Selective out-migration of wealthier residents is yet another 'snowballing' process. The middle classes abandoned British cities in large numbers after the 1970s (see pages 65 and 193 for an account of counterurbanisation). In the USA, the process was dubbed 'white flight' by the media when it left some districts populated mainly by low-income African-Americans. Detroit has lost 1 million residents since 1950, most of them white.

- In turn, the out-migration of professionals left some inner-city hospitals and surgeries with a deficit of well-qualified staff at a time when pressure was mounting on health services (due to rising crime and substance abuse). Education services were similarly affected as inner-city schools lost staff and struggled to attract high-calibre teachers to areas that were increasingly represented in the media as problem places (reiterating the important message from Chapter 2 that place meanings and representations *do* matter and have real consequences for people and places).

🔑 KEY TERMS

Broken windows theory
This theory of place engagement describes the way that environmental deterioration quickly accelerates in locations where there are low levels of active citizenship and place engagement among the local community.

Cycle of deprivation
A vicious circle of linked negative economic, social and environmental changes in any area experiencing stresses such as those associated with deindustrialisation. The cycle may accelerate over time due to positive feedback processes.

▲ **Figure 3.7** A deindustrialised inner-city area where social and environmental conditions have deteriorated uncontrollably: few businesses would want to stay here, even if it were still profitable to do so

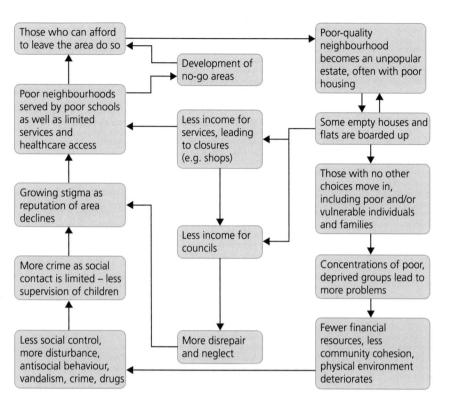

▲ **Figure 3.8** This illustration of the cycle of deprivation shows the operation of multiple connections, interrelationships and feedback loops

Crossing the threshold

A **threshold**, or tipping point, can sometimes be identified when a place undergoes decline. The loss of one particularly vital industry or service can accelerate change. A line is crossed beyond which economic decline is, in effect, irreversible. In a deprived urban neighbourhood, the closure of a local bank or post office (due to dwindling demand for services as local incomes fall) can be the tipping point that causes entire shopping parades to collapse, like toppling a line of dominos. Other local businesses that relied on casual custom from cash-point users find their sales of convenience goods have suddenly dried up. A spate of closures quickly follows. Chapter 6 (see page 194) examines how similar processes may operate in a rural context.

A wicked problem

The rapid onset of deindustrialisation sometimes left an entire generation of school-leavers 'stranded'. Until the 1970s, 15- and 16-year-olds had, for many generations, followed a pathway which led from school directly to serving a traditional industrial apprenticeship, for example in Sheffield's steel industries. Far fewer teenagers than today stayed on at school to take A-levels: in the 1980s, only one-in-eight progressed to university (today, it's more like one-in-three).

 **KEY TERM**

Threshold A critical limit or level in a system or environment which, if crossed, can lead to massive and irreversible change.

The overnight disappearance of traditional industries meant many young school-leavers in the 1970s and 1980s were surprised to find there was suddenly far less manufacturing or mining employment available locally. But they lacked the qualifications needed to work in many service industries or to retrain professionally at a university.

KEY TERM

Wicked problem A challenge that cannot be dealt with easily owing to its scale and/or complexity. Wicked problems arise from the interactions of many different places, people, things, ideas and perspectives within complex and interconnected systems.

Due to its many links and interactions, the cycle of deprivation is an inherently complex challenge – or wicked problem – to tackle. What is the best way therefore to break the tightening circle of entrenched and deepening poverty in deindustrialising areas? What can be done to stem the self-reinforcing outflow of middle-class citizens from the worst-affected places? Chapters 4 and 5 explore strategies to deal with these issues. As you will see, taking action to attract higher-income groups back into inner urban areas is often a priority for city managers (see pages 166–199). This is because of the way it can help generate *virtuous* positive feedback effects (in contrast to the vicious circles analysed in this chapter).

2 Twenty-first-century economic, political and technological challenges

▶ *What are the latest externally-driven challenges for local places in the UK?*

Economic challenges of hyper-connectivity

A series of economic, geopolitical and technological challenges have brought further industrial closures to the UK in recent years. These latest changes have adversely affected places previously viewed as 'successful' or resilient in relation to global shift and deindustrialisation. It is not only the UK's surviving manufacturing industry that is once again threatened; in a hyper-connected world, tertiary and even quaternary industries are at risk too.

The global financial crisis (GFC)

The global financial crisis (GFC) that began in 2007 triggered a slowdown in global economic growth from which the world economy took around a decade to recover. Although the overall value of global trade has expanded enormously in recent decades, its growth rate has slackened notably since the GFC. Some places have lacked the resilience needed to cope with this latest economic storm.

- The GFC originated in US and EU money markets, where sales of high-risk financial services and products triggered the failure or near collapse of several leading banks and institutions. The resulting shockwaves undermined the entire world economy. Some countries experienced severe economic difficulty in the immediate aftermath, including Portugal, Greece and Ireland.

- Several key data indicators show that a cyclical or longer-term downturn in world trade flows has continued to affect developed, emerging and developing economies alike since the GFC (see Figure 3.9).
- International flows of trade, services and finance grew steadily between 1990 and 2007 before collapsing and stagnating. The year 2016 was the fifth consecutive year when global trade did not grow.

The GFC and changing places in the UK

Economists view the period of reduced national output which began in 2008 as the longest economic downturn on record for the UK. The Government's level of debt soared after it was required to 'bail out' two major British banks (Lloyds and the Royal Bank of Scotland). To avoid its debt problem worsening, politicians cut public spending severely in 2010. A series of so-called 'austerity measures' included:

- cuts and freezes in benefits, including housing allowances and child benefit payments
- public-sector cutbacks including pay freezes and redundancies (between 2010 and 2015, 400,000 public-sector jobs were lost)
- reduced arts and culture grants, including a loss of funding for more than 200 arts bodies, following a 30 per cent cut in the budget for Arts Council England.

These austerity measures affected some places more than others. Unemployment rises were greatest in those towns and districts where unusually high numbers of people work in the public sector. In 2013, 500 Liverpool charity workers were made redundant along with 600 NHS workers and 1700 Merseyside fire and police services workers. Places in the borough of Sefton were badly affected by these and other job losses, including Crosby and Maghull.

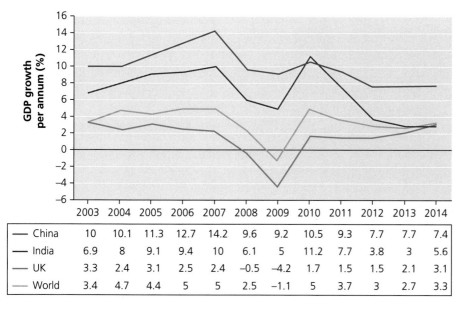

	2003	2004	2005	2006	2007	2008	2009	2010	2011	2012	2013	2014
China	10	10.1	11.3	12.7	14.2	9.6	9.2	10.5	9.3	7.7	7.7	7.4
India	6.9	8	9.1	9.4	10	6.1	5	11.2	7.7	3.8	3	5.6
UK	3.3	2.4	3.1	2.5	2.4	−0.5	−4.2	1.7	1.5	1.5	2.1	3.1
World	3.4	4.7	4.4	5	5	2.5	−1.1	5	3.7	3	2.7	3.3

◀ **Figure 3.9** Annual economic growth 2003–14 for the UK, China and India compared with world average: the immediate impact of the GFC is clearly visible in 2007–09

▲ **Figure 3.10** Blackpool is struggling economically. One view is that the GFC and austerity measures may have pushed places like this to a tipping point

The diverging current state of high streets across the UK gives us insight into the varying resilience of different places following the GFC and the roll-out of austerity measures. Some places have been hit badly by the disappearance or partial closure of retail chains such as Woolworths, Focus DIY, TJ Hughes, Habitat, HMV, Dixons and BHS. These businesses struggled to adapt to harder trading conditions after 2008. One-quarter of all main high streets in the UK were still suffering serious decline in 2017, with many vacant premises left unfilled. The places with the greatest numbers of closures have tended to be those where austerity measures have had the greatest negative impact on the local community's collective purchasing power, such as Castleford and Blackpool (see Figure 3.10).

Places with the worst-performing high streets have been dubbed 'terminal towns' by the retail analyst Colliers International. These settlements are struggling to find either public or private finance to spur future growth. One view is that some UK towns have already reached a tipping point that has sent them into a downward spiral from which recovery seems unlikely: with sales in shops falling year-on-year as people turn to online shopping (see page 96), conditions are unlikely to improve soon.

Competition from emerging economies

Both in the run-up to the GFC and during its aftermath, continued strong growth for many of the world's major emerging economies has been another important factor affecting British industries. Pressures on UK manufacturers in the 1960s and 1970s came mainly from an earlier wave of Asian industrialisation in Japan, South Korea, Taiwan, Singapore and Hong Kong. More recently, British workers have found themselves competing with industries and employees in China, India, Indonesia, Bangladesh and other emerging economies. Moreover, India and China have become increasingly important providers of tertiary and quaternary services in addition to manufacturing. This has had implications for all industrial sectors in the UK and places that rely on this employment, as Table 3.3 shows.

However, it is important to recognise that emerging economies create new opportunities, not just challenges, for places in the UK. Shifting flows of investment operate in *both* directions. The Port Talbot steelworks in South Wales would not have stayed open without inward investment from India. Chinese money from sovereign wealth funds has bankrolled numerous large infrastructure projects in the UK, bringing much-needed new employment to some local communities (see page 128), although another view is that it leaves the UK beholden to external powers. Finally, it is worth reflecting on the ways in which innovation and the mass production of

🔑 **KEY TERM**

Sovereign wealth funds (SWFs) Government-owned investment funds and banks, typically associated with China and countries that have large revenues from oil, like Qatar.

electronic devices in emerging economies have played an important role providing people throughout the UK with the smartphones that have utterly transformed how we interact with one another and places that surround us.

Sector	How the growth of emerging economies has affected places in the UK
Secondary industries	Manufacturing employment in the UK has continued to decline, falling by a half between 1991 and 2011 (see Figure 3.2). This is explained in part by the emergence of China, Indonesia and, more recently, Bangladesh and Vietnam (among many others) as highly attractive destinations for footloose global manufacturing capital. Many British companies, including Marks and Spencer and Tesco, now outsource clothing production to Bangladesh. Hornby – the maker of model trains and Scalextric – has made its toys exclusively in China since 2002 while iconic English pottery brands, including Wedgwood and Royal Doulton, moved production to Bangladesh and Indonesia in an attempt to remain profitable. In such cases, British-based factory closures brought unemployment to places where families had worked in these industries for generations.
Tertiary industries	Around 2000, some UK-based service providers, including major banks and utility companies, began moving their call-centre operations to India or the Philippines. Historical place connections help explain these movements, along with significantly cheaper labour costs: English is widely-spoken in India (a legacy of British Imperialism), making cities like Bangalore a suitable site for call centres providing support services to British customers.
Quaternary industries	Even quaternary industries are not immune to this latest round of global shift. In 2017, China registered more than 1,380,000 new patents. This represented a 3000 per cent increase since 2000, putting China ahead of the United States in the new knowledge-creation race. Increasingly, UK research hubs, including technology clusters in Bristol, Cambridge and other 'smart cities', must compete with China for business.

▲ **Table 3.3** Economic growth in China, India and other emerging economies has brought further UK employment losses spanning multiple industrial sectors since 2000

CONTEMPORARY CASE STUDY: INDIA, CHINA AND THE PORT TALBOT STEELWORKS

Few places illustrate the challenges of interconnectivity better than Port Talbot in South Wales. Here, the fate of up to 15,000 workers – whose livelihoods depend directly or indirectly on Port Talbot steelworks (see Figure 3.11) – continues to hinge on the outcome of evolving global interactions between India, China, the UK and the European Union (with all the current uncertainty that entails).

▲ **Figure 3.11** The steelworks skyline of Port Talbot

During the nineteenth century, South Wales mainly benefited from its global connections. The region grew into an important global hub for steel production during the 'golden age' of the British Empire. This success did not last though. By the 1990s, only two large integrated steelworks remained in South Wales following the global shift of steel manufacturing first to South Korea and later to China. One of these two survivors was the Port Talbot steelworks which opened in 1954 with many locational advantages, notably its proximity to the M4 motorway corridor.

Ownership of the Port Talbot steelworks passed from British Steel to the merged Anglo-Dutch group Corus in 1999 and then to Indian transnational corporation Tata Steel, following a £7 billion takeover in 2007. The changing geography of the steelwork's ownership during this time period was part of a far bigger historical picture whose elements included the UK Government's sale of its own industrial assets into private hands and the growing global influence of Asian states and businesses. At the time, commentators saw historical irony in the acquisition of the remnants of British Steel by an Indian company: several newspapers ran the story under the *Star Wars*-inspired headline: 'The Empire Strikes Back'.

For two reasons, Port Talbot failed to meet Tata Steel's initial profit expectations.

- Worldwide demand for steel fell after the GFC of 2007–09.
- China began producing a surplus of steel soon after the GFC, which it dumped on world markets, including Europe. Benefiting from low labour costs and government subsidies, Chinese-made steel was far cheaper to buy than anything produced in Port Talbot.

Several European governments including Italy demanded the EU retaliate against unfair Chinese steel prices with import tariffs. But many other EU states were against any action being taken. In particular, the UK Government under David Cameron was firmly opposed to any proposal of tariffs being imposed on Chinese steel, despite the clear implications of inaction for Port Talbot. Unfortunately for South Wales, the UK Government's wider geopolitical strategy was to strengthen economic and political ties with modern China, which meant it wanted no part in a potential trade war. There was no desire among British politicians to jeopardise China's role as an increasingly important funder of large existing and proposed infrastructure projects in the UK, such as the Hinkley Point nuclear plant in Somerset and the High Speed 2 railway.

Unchecked Chinese imports meant that by 2016 Port Talbot steelworks was reportedly losing £1 million a day. Finally, an announcement was made by Tata Steel that 1000 jobs would be cut prior to the steelworks being sold or shut down altogether. The threat of closure put 4000 jobs directly at risk plus an estimated further 11,000 tied to local supply chains and support industries. In a press statement, Tata Steel blamed its decision on the flood of 'unfairly traded' cheap imports, 'particularly from China', and criticised the European Union for not taking measures to protect the future of Port Talbot and other steel-making places.

During most of 2017, the future of Port Talbot remained at risk until a rescue package was agreed finally by Tata Steel and the UK Government.

- Tata Steel made a new five-year commitment to keep both of Port Talbot's blast furnaces open until 2021.
- In return, the UK Government allowed the company to close an expensive and generous pension scheme for local workers. The UK's ageing population had turned this pension fund into a massive financial liability for Tata Steel. In the future, new workers will be given less generous pension benefits.
- Tata Steel's decision to continue operations in Port Talbot coincided with a partial recovery in sales of British steel because of the fall in the value of the pound after the Brexit referendum: thanks to external events, Port Talbot steel had become cheaper to buy.

However, there is uncertainty about what will happen beyond 2021, particularly given the announcement by Tata Steel and Germany's Thyssenkrupp of an agreement to combine their European operations in a joint venture. The result could be rationalisation and further job losses at Port Talbot after 2021. There are also fears about what else might happen to the value of local people's pensions.

In conclusion, this case study's 'big story' is the tug of war between two emerging superpowers, India and China, with Port Talbot steelworks stuck in the middle. The future of this place remains highly uncertain because of complex political and economic forces interacting at local, national, international and global scales (see Figure 3.12).

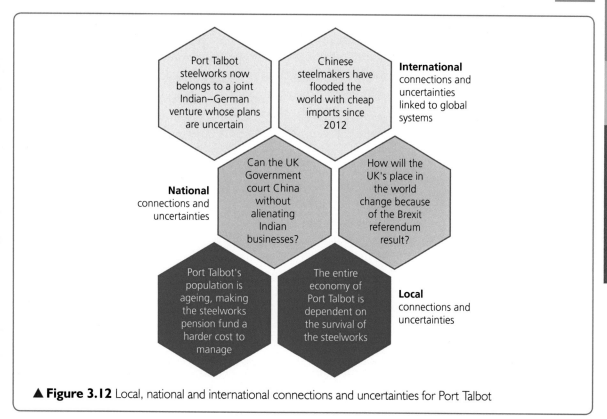

▲ **Figure 3.12** Local, national and international connections and uncertainties for Port Talbot

Political changes and challenges for places

Political factors help determine the prosperity (or lack of it) of different places. Countless past examples serve to illustrate how the fate of local communities and high streets can hinge on planning rules, government legislation and national membership of international organisations. Consider briefly, for example, how the fortunes of different towns and cities were affected by:

- the political decision to build motorways and new towns in the 1950s
- the introduction of Sunday trading in 1994 and changes in licensing laws allowing bars to stay open later
- large numbers of migrants arriving in the UK after Poland and other eastern European countries joined the EU in 2004.

Two contemporary illustrations of local impacts caused by political decision-making are explored next. These are (i) the UK's planned exit from the EU and (ii) the drive towards a low-carbon (or 'decarbonised') economy.

Implications for places following Brexit

Around 52 per cent of those who voted chose to leave the European Union in the 2016 referendum on membership. So great is the complexity of arranging a 'divorce' from the EU that many of the details about what is actually going to happen will remain unclear until 2019 or later. Whatever the exact outcome, it is reasonable to expect major changes in the size and direction of those flows of goods, investment and people that currently link places in the UK with Europe and the wider world. Two key issues must be resolved (and may have been by the time you read this).

● *Will the UK remain part of the European single market which allows free movement of goods, services, money and people within the European Union?* The single market is seen by its advocates as the EU's biggest achievement and is one main reason why the UK joined in the first place. Many UK service industries (including insurance companies and banks) have benefited from 'frictionless' single-market arrangements. But many UK citizens became increasingly unhappy with the unchecked free movement of people into the UK.

● *Will the UK remain part of the European customs union?* This is a separate issue from the single market; staying in the customs union limits the UK's ability to strike its own trade deals with non-EU countries.

Whatever happens, unpacking 43 years of laws, treaties and agreements covering thousands of subjects will be a very complicated task. The full implications of Brexit for different places around the UK will only become clear in the fullness of time. Table 3.4 provides a summary of some important potential issues.

Area of concern	Main issues
Customs and trade	Businesses that rely almost exclusively on exports of goods and services to the EU may suffer a fall in sales. In a worst-case scenario, this might even lead to further deindustrialisation in some localities. However, increased costs of trade may be offset by other impacts such as the changing value of the pound. As a result, some people expect that British businesses – and the places they support – will be resilient.
Supply chains	Many industries, such as UK car makers, rely on imports of many different components. Prices may increase for the parts they need, which could drive some companies out of business or force them to relocate to another EU country. However, optimists say that Brexit may encourage more UK entrepreneurs to set up new businesses supplying parts to the companies that need them: this could create positive multiplier effects and new employment in places where work is needed.
Workforce	An end to the free movement of people could leave many farms, construction sites and food-processing companies without the plentiful cheap supply of eastern European migrant workers they have grown reliant on. However, the opposing view is that there will be more jobs available for UK citizens in the future, and that 'guest worker' passes could always be issued to temporary migrants if a need arises.
Foreign direct investment (FDI)	It remains unclear whether the UK will still be seen as a favourable destination for FDI from TNCs based in the USA, China, India, Japan and other non-EU countries. Some foreign carmakers, such as Nissan in Sunderland, have said they will stay in the UK post-Brexit. Other TNCs are doubtless waiting to see what happens. It is possible that some TNCs will be deterred from locating in the UK if it leaves the single market as this could increase trading costs. However, there are plenty of good reasons why TNCs may want to retain a presence in the UK, including the size of the British market and the value of their existing investments.
Regulatory changes	An exit from the EU will mean that the UK may choose to adopt different environmental rules in the future: this has all kinds of implications for places and landscapes.

▲ **Table 3.4** Some possible impacts of Brexit for places, people, businesses and landscapes in the UK

Implications for places of a low-carbon economy

The political drive to decarbonise Britain's economy began in earnest around the time of the Stern Review (2006), which recommended that global energy production must become significantly decarbonised by 2050 if atmospheric concentrations are to stabilise at or below the critical threshold level of around 550 ppm CO_2. The Climate Change Act 2008 set a 'legally binding' national emission reduction target of 80 per cent by 2050 (measured against a 1990 baseline).

The subsequent promotion of cleaner energy sources has already left its mark on many landscapes and communities, as outlined below.

▲ **Figure 3.13** In 2016, a small majority of UK voters chose to leave the EU

- Coal demand continues to decline; in April 2017, the share of electricity from the country's shrinking number of coal power plants fell to zero for the first time in 130 years. In the run-up to the 2015 adoption of the Paris climate change accord, the UK Government said it wanted to phase out coal power entirely by 2025.
- Wind turbines have changed the appearance of landscapes throughout the UK, often becoming a focus for local protest. Bradford Council's decision to approve a wind power project for Haworth in Yorkshire – the setting for Emily Brontë's famous 1848 novel *Wuthering Heights* – is explored on page 202.
- Fracking for gas is broadly supported by the UK Government on account of the lower carbon footprint of gas compared with coal, in addition to the contribution new domestic gas supplies can make to national energy security. Fracking involves injecting water into the ground to fracture the rock and free the gas. In 2014, people protested in the Sussex village of Balcombe, where an energy company, Cuadrilla, had begun exploring for gas. Some superglued themselves to the window of London offices used by Cuadrilla.
- Despite concerns after Japan's Fukushima nuclear power station disaster in 2011, the UK Government approved the construction of a new nuclear power station at Hinkley Point in Somerset. EDF Energy asserts that 25,000 new jobs will be created; posters designed by residents show their concerns about the new Hinkley Point C power station (these posters can be viewed at their website, stophinkley.org).
- The £1.5 billion, 320 megawatt Swansea tidal power project originally won planning permission in 2015 despite environmental objections, but doubts over the cost-effectiveness of the project have hindered progress.

ANALYSIS AND INTERPRETATION

Study Figure 3.14, which shows coal use in the UK since 1860.

(a) Using Figure 3.14, analyse changes over time in the amount of coal used in the UK.

> ## GUIDANCE
>
> The command word 'analyse' requires a well-structured answer that draws out the most essential elements of a graph or table. It is good practice to begin by establishing what the 'big picture' or 'headline' is – in this case, it is a 'rise and fall' story (although the initial rise was far smoother than what followed). There are, of course, important short-term fluctuations to identify and include in the analysis: the three most marked short-term falls in coal use are all attributed to strikes. Perhaps the greatest challenge with an analysis of this kind of visual resource is knowing what to leave out. Some fluctuations may not really be worth mentioning when taking a longer-term view, while some of the annotations on the graph are perhaps more worthy of mention than others. If you were to restrict your answer to just 100 words, what would be the most important points to include in your analysis?

(b) Suggest how coal-mining places and their communities may have been affected by the changes shown in Figure 3.14.

> ## GUIDANCE
>
> This is a relatively open-ended question with few restrictions on what you can or cannot write about. As a result, it is a good idea to produce a plan before starting to respond. You could structure an answer by providing a series of paragraphs which deal sequentially with:
>
> - the environmental impact on coal-mining regions of increased production until the 1960s and decreases in output which followed
> - population growth and migration trends for coal-mining places during the growth and decline phases shown in Figure 3.14
> - the growth of a sense of community during the boom years and ways in which this could have been represented to the world (such as the musical or sporting life of coal-mining communities), and the subsequent loss of community cohesion and fragmentation of place identity in the years following the coal miners' strike of 1984.

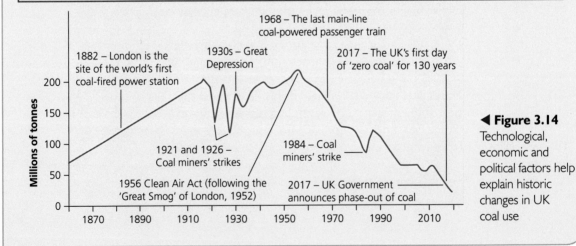

◀ **Figure 3.14**
Technological, economic and political factors help explain historic changes in UK coal use

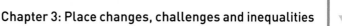
Technological changes and challenges for places

New technology both creates and destroys jobs: inevitably there are winners and losers, both in terms of people and places. As the Industrial Revolution progressed, old jobs were swept away by increasingly sophisticated automated machinery. Fierce market competition created perfect conditions for continued innovation as firms vied with one another to reduce production costs and maximise sales. But new employment arose too as a result of innovation: new products need people to build, market and sell them.

The same forces are still at play today, though the technology differs. Table 3.5 provides a glimpse of emerging threats to places posed by online retailing, artificial intelligence, advanced robotics and 3D printing.

Technology	Threat to places
Online retail	■ In some small market towns which lack a diverse economy, rising competition from online retailing is expected to lead to more shop closures and possible 'tipping point' effects in coming years. ■ Even some of the more resilient shopping centres could experience rising unemployment: Amazon opened its first supermarket without check-outs in January 2018 in Seattle in the USA. Customers scan their Amazon Go app and goods are charged to their account as they are removed from shelves.
Artificial intelligence (AI) and robotics	■ In 2017, the Bank of England stated that many millions of British jobs are threatened by AI. 'Robo-advisers' could lead to the loss of traditional high-street travel agents; driverless cars could leave taxi drivers unable to earn a living; many more sectors of industry are braced for change. ■ Jobs most at threat from the next wave of automation are those requiring a low skill set. Some Fenland towns, near Cambridge, currently enjoy high employment in food-processing work. However, these jobs are likely to be replaced by machines. ■ According to a 2018 Centre for Cities report, jobs that are made up of routine tasks are at a greater risk of decline, whereas those occupations requiring interpersonal and cognitive skills are set to grow. In Mansfield, Sunderland, Wakefield and Stoke, almost 30 per cent of the current workforce is in an occupation that is very likely to shrink by 2030. This contrasts with cities such as Cambridge and Oxford, where fewer than 15 per cent of jobs are thought to be at risk.
3D printing	■ When we paint on paper, thick layers can be built up with repeated brushstrokes. 3D printing takes the same approach by 'painting' layer upon layer of a resin or polymer, until fully three-dimensional objects are created. Some specialised items are now produced this way: fuel nozzles have been printed for Boeing and Airbus aircraft, for example. ■ Although the disruption of mass-produced goods, such as kettles and toasters, is still a long way off, it is clear that future manufacturing of some everyday items will, potentially, take place anywhere, allowing customers to print much of what they need rather than order it from businesses. How could this affect some industries and those places that depend on them for employment?

▲ **Table 3.5** New technology threatens certain jobs and, in turn, those places that depend most on these types of employment

In late 2017, 18 months after the Brexit referendum, data showed large numbers of European migrant workers beginning to find employment in markets other than the UK. Pro-Brexit politicians have argued this will increase availability of employment in some places for British workers; however, nothing can be taken for granted given the rapid pace with which artificial intelligence and robotics are now advancing. Some experts view the loss of cheap European workers as a possible catalyst for more British employers to invest in further automation.

The rise of Amazon is an interesting theme to explore further as part of your Changing Places studies.

- E-commerce has changed many people's experience of living in remote rural areas: in the past, communities may have felt an exaggerated sense of isolation because of inadequate access to retail services. In contrast today, goods can be ordered online at the click of a button and are delivered the next day.
- Amazon deliveries are beginning to affect urban people's experiences of living in high-density areas because of the additional road traffic congestion that e-commerce deliveries are generating. Amazon's response to this has been to take its first steps towards delivering goods using unmanned aerial vehicles, or drones. In the future, drone deliveries could become part of everyday life in the place where you live.

CONTEMPORARY CASE STUDY: THE LOSS OF BHS DEPARTMENT STORES FROM UK HIGH STREETS

The chain store BHS (British Home Stores) collapsed in 2016. As a result, a large gap opened in 164 town and city centres across the UK where BHS closed stores. Due to the size of BHS department stores, their loss had a large and immediate impact on employment and retail footfall in the affected areas, much as the collapse of chain store Woolworths had done previously in 2009. For some provincial shopping streets, BHS clothing and home supplies had been their biggest attraction.

▲ **Figure 3.15** BHS department stores closed their doors in 2016

■ It is important if you are carrying out an analysis of shop closures to acknowledge that the loss of a large department store has a far more damaging impact than the loss of a chip shop.

■ Removing a large department store can be akin to removing a vitally important brick from the bottom of a 'Jenga' tower: it's the tipping point of no return which leads to the collapse of the entire structure.

The loss of BHS was felt acutely by neighbouring retailers who relied on it for their own footfall. As one retail analyst notes: 'high streets are delicate ecosystems that can be disturbed when a long-established business is felled'. For example, more than half of Ann Summers' UK stores are within 200m of a former BHS site, leaving the lingerie company disproportionately exposed to BHS's failure.

One year after BHS closed its doors, very few of the sites had been taken over by an alternative department store operator. The reason for this was the sheer size of many stores. Discount clothing store Primark adopted thirteen of the properties from Carlisle to Llandudno. Others have become discount supermarkets. In a few cases, the ground floor has been taken over by a retailer while the upper levels have been converted into offices.

A major reason for the failure of BHS is the rise of online retailing and Amazon in particular, which now makes up around 20 per cent of all UK sales.

Challenges or opportunities?

Winds of change can bring devastation to places but a chance for renewal too; the technological, economic and political processes outlined above have brought new challenges, to be sure, but there are also fresh opportunities for development and innovation in response. With hindsight, the broad transformation of the UK into a post-industrial economy and society – necessitated by the immovable forces of global shift and deindustrialisation – was a welcome change that brought environmental improvement and life expectancy gains. Figure 3.16 shows one view of this transition.

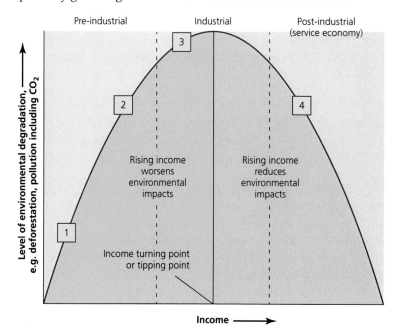

◀ **Figure 3.16** The environmental Kuznets curve portrays UK deindustrialisation in a positive light (while demonstrating also how global shift has adversely affected the environments of other globally-connected countries such as Bangladesh and China)

Key
1. Europe pre-Industrial Revolution, remote Amazonia today, Bangladesh pre-1970s
2. Bangladesh today, China in the twentieth century
3. China today
4. Developed countries today

Figure 3.17 shows how external shocks can prompt resilient place remaking in ways that may ultimately lead to a better state of affairs than before. Resilient places can adopt new technologies and seize opportunities to adapt and innovate economically in ways that ultimately bring an improved economy and enhanced quality of life for the local community. Chapters four and five chapters explore ways in which this can be done.

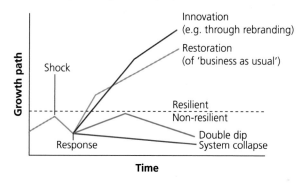

◀ **Figure 3.17** Resilience means having the capacity to leap back or rebound, following a disruption or disaster; the illustration shows how the concept of resilience is applied to an economy's response to a financial crisis

③ Changing demographic and cultural characteristics of places

▶ *What changes and challenges have increased cultural and demographic diversity brought to different places?*

The demographic and cultural geography of the UK has undergone a transformation in recent decades in parallel with the economic developments outlined above. While industrial landscapes were being reshaped by outflows and inflows of global investment, the demographic and cultural characteristics of places were also changing because of shifting flows of people and ideas.

Demographic processes and place changes

You are probably acquainted with the main ways in which fertility, mortality, life expectancy and population structure have changed over time in the UK. Table 3.6 provides a summary of the main points to remember.

Variable	Changes over time
Population size	■ Total population in the UK has crept upwards from 58.8 million in 2001 to 66 million in 2016. The increase is attributable to people living longer and migration into the country. ■ Growth in numbers has not occurred in all places, however. There has been rapid growth in London and the southeast. London's population has grown from under 7 million in the 1980s to approaching 9 million today. Slower growth is seen in depopulated regions of mid-Wales and western Scotland, and growth has also been slow in large tracts of northern England.
Vital rates	■ The fertility rate is the average number of children a woman gives birth to in her lifetime. It has fallen to below 2.0 in the UK over time on account of the emancipation and professionalisation of women. This story encompasses the women's suffrage movement of the 1910s, the right to choose (abortion was legalised in 1967) and equal pay legislation in 1970. ■ Generally, mortality in the UK has fallen because people are more aware of their vulnerability to a range of risks (thanks to guidance on nutrition, including recommended limits for calories, saturated fat, alcohol and salt intake). However, the wealth and lifestyle choices of different communities mean that mortality can vary greatly between different places and postcodes.
Population structure	■ The UK has an ageing population. Continuing improvements in healthcare mean that the number of people aged 85 and over reached a record 1.4 million in 2017. This is the equivalent of one-in-every-50 people. ■ The population of over 85s grew from 1.1m in 2001 to 1.4m in the 2011 Census (this is the fastest growth rate for any age group in the UK). By 2066, there are predicted to be over half a million people aged 100 or over in the UK. There are now more than 10 million people aged 65 and over.

▲ **Table 3.6** Demographic changes in the UK

For the purposes of this book, it is important to think critically about, first, impacts of the developments shown in Table 3.6 on different places in the UK and, second, the role that flows of ideas and people have played in bringing about these demographic changes.

Spatial variations in demographic changes

The impacts of population changes vary enormously from place to place in the UK. Figure 3.18 shows the enormous disparity in life expectancy that exists between different local authority areas. Longevity varies greatly both between and within places: behind the average UK figures of 77 years for men and 82 years for women lies a wide range of local life expectancy data. In South Cambridgeshire, men can expect to live around 10 years longer than those in Glasgow. Income, occupation and education are key factors explaining these differences, along with lifestyle choices such as diet and smoking.

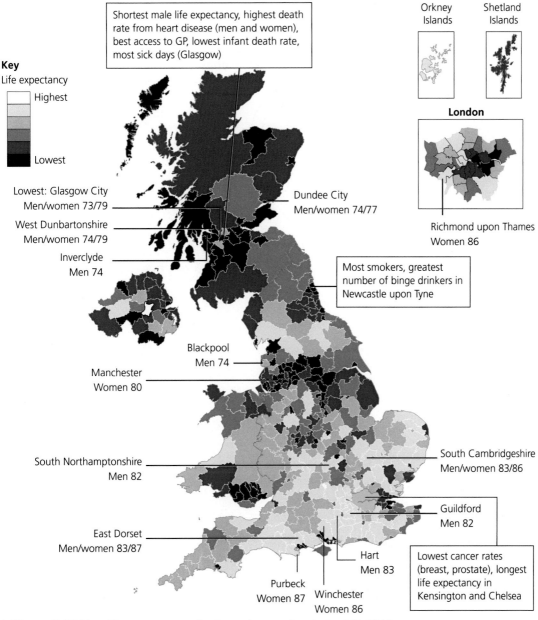

Orkney Islands   Shetland Islands

Shortest male life expectancy, highest death rate from heart disease (men and women), best access to GP, lowest infant death rate, most sick days (Glasgow)

Key
Life expectancy
Highest
Lowest

London

Lowest: Glasgow City
Men/women 73/79

West Dunbartonshire
Men/women 74/79

Inverclyde
Men 74

Dundee City
Men/women 74/77

Richmond upon Thames
Women 86

Most smokers, greatest number of binge drinkers in Newcastle upon Tyne

Blackpool
Men 74

Manchester
Women 80

South Northamptonshire
Men 82

South Cambridgeshire
Men/women 83/86

Guildford
Men 82

East Dorset
Men/women 83/87

Hart
Men 83

Lowest cancer rates (breast, prostate), longest life expectancy in Kensington and Chelsea

Purbeck
Women 87

Winchester
Women 86

▲ **Figure 3.18** How life expectancy varies from place to place in the UK, 2011

Significant differences exist too in the proportion of older people living in different parts of the UK. Variations can be identified at varying scales: some counties and local authorities are far 'greyer' than others; local neighbourhoods within the same settlement can also exhibit marked differences in the proportion of people aged 65 and over for a range of reasons linked to the varying residential preferences of different working and non-working age groups.

Figure 3.19 shows regional variations in the proportion of the population aged 65 and over. The differences can partly be explained by variations in life expectancy but are also the result of age-selective migration processes (large numbers of elderly people move to coastal areas for their retirement, while centripetal forces instead help draw younger people towards metropolitan areas where universities and employment are found). Many coastal settlements such as Worthing or Southport have high proportions of older people living there as a result of age-selective migration.

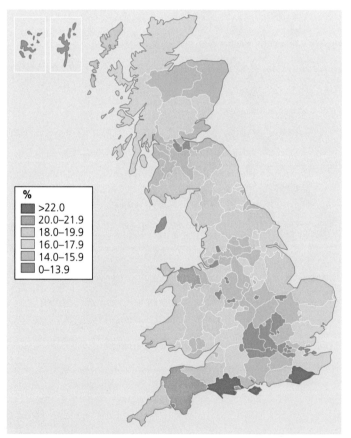

%
| >22.0 |
| 20.0–21.9 |
| 18.0–19.9 |
| 16.0–17.9 |
| 14.0–15.9 |
| 0–13.9 |

▲ **Figure 3.19** Regional variations in the proportion of the population aged 65 and over, 2011

Variations in population structure affect the characteristics of local places in many ways and also help determine how they are perceived by outsiders.

- You can often judge from the retail make-up of a high street whether it serves a more elderly or youthful population (think what kinds of differences you might expect to see).
- Levels of community engagement may differ for places with a high proportion of retirees as opposed to, say, a higher proportion of students (retired people are more likely to be active members of local societies and charities, for example).
- Some places with an ageing population are being physically modified in ways that reflect their changing demographic characteristics. Some local supermarkets cater for elderly shoppers by including brighter lighting, non-slip floors and extra-wide aisles that mobility scooters can navigate easily. Outside of the UK, more examples can be found of places that have been remodelled specifically for elderly populations. For example, a specially-built neighbourhood, Hogeweyk, is located in the Dutch town of Weesp. It is notable because it has been designed specifically as a pioneering care facility for elderly people with dementia.

Demographic place changes in a global context

The demographic changes affecting communities and settlements throughout the UK have, of course, been shaped by ongoing national and global developments in healthcare. This is another example of the way in which exogenous factors (see page 21) continually exert influence over the local places we study. Numerous 'global' reasons help explain why large numbers of people now survive into their eighties and nineties.

- Nutrition has improved, thanks in part to the global food production networks of large supermarkets that supply fresh fruit and vegetables all-year-round to stores in every British town and city.
- Disease prevention and treatment have improved worldwide thanks to the global collaboration of researchers and scientists, often under the auspices of international agencies such as the World Health Organization (WHO). For example, the last outbreak of smallpox happened in 1978 (shortly prior to its eradication in 1980 following a global immunisation campaign led by the WHO). There hasn't been a case of polio in the UK since the 1990s and this is due in large part to a sustained European effort of immunisation and disease surveillance.
- There are countless further ways in which interconnected global communities of scientific researchers have, over time, generated knowledge and understanding which has helped local communities everywhere to mitigate life-threatening risks. The fact that we now know far more about the dangers of smoking is a very important reason why British people live longer than they used to (walking into a pub in any place in the UK as recently as the 1990s, you might have encountered a thick cloud of cigarette smoke).

ANALYSIS AND INTERPRETATION

Figure 3.20 is a socioeconomic profile of Blackpool, a declining coastal resort town in northwest England. In this profile, Blackpool's life expectancy is compared with other UK local authorities. Net inflows and outflows of people per year are also shown for Blackpool.

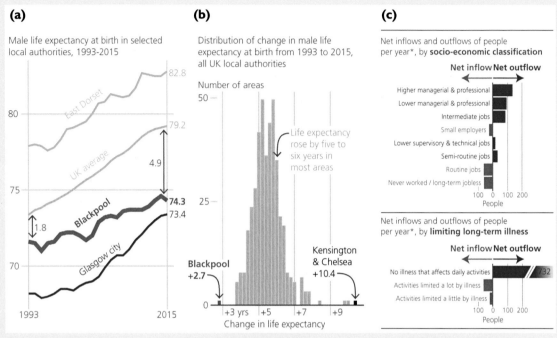

(a) Male life expectancy at birth in selected local authorities, 1993-2015

(b) Distribution of change in male life expectancy at birth from 1993 to 2015, all UK local authorities

(c) Net inflows and outflows of people per year*, by **socio-economic classification**

Net inflows and outflows of people per year*, by **limiting long-term illness**

▲ **Figure 3.20** A socioeconomic profile of Blackpool, 2015

(a) Using Figure 3.20a, calculate the increase in UK average life expectancy between 1993 and 2015.

(b) Using Figures 3.20a and 3.20b, compare life expectancy changes for Blackpool with other areas of the UK.

GUIDANCE

There are two figures to make use of in your answer. Figure 3.20b is useful because it shows inequality in the UK as a whole and reveals that Blackpool is the worst-performing of all local authorities. No other area has an increase of less than 3 years as Blackpool does; the best-performing area, Kensington and Chelsea, has an increase almost four times higher than Blackpool's. There are two modal frequencies to identify and the value for Blackpool is significantly lower than these. Figure 3.20a provides a more detailed insight into how Blackpool compares geographically with another poorer northern urban area (Glasgow City) and also an affluent southern rural area (East Dorset). Figure 3.20a additionally provides the average UK data. This shows us Blackpool's worsening life expectancy over time in comparison with the UK as a whole. The overall 'big story' here is that although life expectancy has increased in Blackpool over time, the gains are poor when compared with all other places.

(c) Explain why Blackpool's decline might begin to accelerate because of the flows of people shown in Figure 3.20c.

GUIDANCE

Figure 3.20c shows the selective net out-migration of more skilled people and the in-migration of unskilled and unemployed people. Also, more people are leaving than arriving overall. The likely result is a decline in both population numbers and the average purchasing power of the remaining population. This, in turn, may lead to the failure of some local shops and businesses because they will sell progressively fewer products before eventually crossing their profitability threshold. It is important to note that the question asks why decline might *accelerate* because of flows of people. This requires us to think of reasons for a possible 'snowballing' effect wherein the decline actually speeds up over time. Why might this be the case? Perhaps the loss of affluent people and the subsequent failure of certain businesses could start to have further knock-on or 'domino' effects that lead to spiralling decline (this is an example of a positive feedback effect in human geography).

(d) Assess the strengths and weaknesses of this socioeconomic profile of Blackpool.

GUIDANCE

This question is relatively open-ended and allows a variety of interpretations of strengths and weaknesses to be written about. First, an assessment can be made of the value of the data chosen compared with other possible indicators of socioeconomic health that might have been used instead. For example, Figure 3.20 focuses heavily on life expectancy at the expense of other socioeconomic indicators. Might it have been more useful to replace either Figure 3.20a or 3.20b with data showing vacant shops, for example? Second, an assessment can be made of the strengths and weaknesses of the presentation methods used. Are all the graphs clear and easy to understand? Is the 'magnification' of y-axis data in Figure 3.20a helpful or misleading in terms of the way it makes small differences appear much larger? Is the labelling of Figure 3.20c sufficiently helpful or would it have been better to include actual income figures for different groups?

The changing cultural characteristics of place

The 2011 Census showed the UK to be a more culturally diverse place than ever before (see Table 3.7). Fewer than 80 per cent of the population viewed themselves as 'White British'. Around one-fifth – around 13 million people – declared they were Asian/Asian British, Black/Black British or White but not born in the UK – a category that in 2011 included millions of EU citizens residing in the UK.

Minority ethnic groups in UK	Percentage of total	
	2001 Census	2011 Census
Black or Black British	2.2	3.4
Asian or Asian British	4.8	7.5
White but not born in UK	3.8	5.3
Multiple identity	1.4	2.2

▲ **Table 3.7** The main UK minority ethnic groups at the time of the 2011 UK Census

This represents a considerable change over time, given that the total size of the UK's minority ethnic population is estimated to have been just one-quarter of a million people in 1951. The changes have been even more pronounced at the local level: populations in parts of rural Wales remain 99.9 per cent White; those of some London boroughs are 60 per cent non-White British (and one-third of Londoners are foreign-born).

Although in general urban places tend to be most culturally diverse, international migration into rural areas has brought demographic and cultural change to these areas, too.

- Rural farming areas surrounding Peterborough have attracted young Polish male migrant workers. In the 1980s, many Portuguese workers moved to farming areas in Lancashire.
- Hotels in the rural island of Arran, Scotland, increasingly rely on young female eastern European staff.

As a result, many rural areas have increasingly culturally diverse populations (although they are still viewed as predominantly 'white' areas, highlighting the difference between ethnicity and race as markers of cultural difference).

Increased diversity can change the cultural landscape of places in various ways. Chapter 2 introduced the idea of cultural traits, such as a community's language, food, religion, clothing and music. Refer back to page 42–44 and think about the various ways in which the landscape of a place could change on account of increased ethnic diversity. It is important to remember, however, that processes of change may occur in two contrasting ways resulting in either homogeneous or heterogeneous new cultural neighbourhoods.

Homogenous places

A particular place or neighbourhood may undergo a cultural change involving the in-movement of a single new **ethnic group** in large numbers. The result is a changed place but one that remains essentially homogenous (uniform in ethnic character). In the 1930s, the American geographer Burgess observed this happening in Chicago: his 'ecological' model of urban land use was based on a theory of 'succession' in which the arrival of a new ethnic group in large numbers would prompt the out-migration of existing people from an inner-city neighbourhood in a process of 'succession'. After a generation or two, the new arrivals would themselves become wealthier and migrate to a better neighbourhood, freeing up the inner-city location for yet another new migrant community. By way of example, Burgess's original model included a neighbourhood called 'Little Sicily' (reflecting its predominantly Italian–American character at that time).

Homogenised ethnic neighbourhoods are sometimes described positively as ethnoscapes and pejoratively as 'ghettos'. Whatever the label, there are places where segregation is evident: a group with a distinctive identity has chosen to live together. Reasons may include:

- the deliberate preservation of cultural heritage and a sense of identity
- functional needs: specialist food services (halal, kosher) and places of worship (mosque, synagogue) operate like any other service which needs its target audience living close at hand.

🔑 **KEY TERMS**

Ethnic group A section of the population which differs from the majority according to criteria such as religion, language, nationality or race. The presence of a minority ethnic group may stem from a past or continuing migration stream. Ethnic groups often exhibit some degree of segregation (separation).

Ethnoscape A cultural landscape constructed by a minority ethnic group, such as a migrant population. Their culture is clearly reflected in the way they have remade the place where they live.

Segregation can also arise as a result of constraints (this is called 'ghettoisation'). For instance:

- discrimination in the housing market (overt discrimination is illegal but 'gatekeeping' still occurs in the rented property market)
- discrimination in the labour market (if a group fails to secure high paid work then this limits their choice of residential areas to live in).

Heterogeneous places

Alternatively, some places may become heterogeneous neighbourhoods composed of any number of different ethnic groups (see Figure 3.21). Some neighbourhoods in London are home to communities drawn from almost every country in the world, from Albania to Zimbabwe; churches, synagogues and mosques share the skyline. The US cartoon series *The Simpsons* is a representation of a heterogeneous community: a diverse cast of characters with multiple ethnic identities populate the fictional settlement of Springfield.

History reveals, however, that levels of diversity at the local level may lessen as time passes. For instance, around 70 per cent of people in London identity themselves as being White British in the 2011 Census. Yet this community was originally far from culturally homogenous. At different times in the past, varied white ethnic communities of Viking, Anglo-Saxon, Celtic, Roman, Viking and Norman descent have all lived in London. Over time, these diverse migrant groups combined in a cultural melting pot that gave rise to modern English culture and language as it is spoken today. It may be the case that, in the future, cultural diversity will begin to lessen in world cities like London, Toronto and Paris that are currently incredibly diverse. This will be on account of cultural intermixing and intermarriage between the different migrant communities.

An interesting account entitled 'Changing Faces' in *National Geographic* (2013) suggested that more and more people in mixed US ethnic neighbourhoods are describing themselves as being of mixed heritage. The inference is that these societies may, over time, cease to be truly diverse. An alternative possibility is that different groups will preserve their heritage; intermixing and intermarriage may be less than expected. As a result, world cities may experience superdiversity (as continued in-migration brings even greater numbers of different nationalities and ethnicities to these places).

▲ **Figure 3.21** A shop's phone card or travel display might provide field evidence that you are in an ethnically heterogeneous place: it could suggest people of many different nationalities live in this neighbourhood (in this case multicultural El Raval, in central Barcelona)

 KEY TERMS

Melting pot A cultural process which involves different communities combining over time (at school, in the workplace or through marriage) to form a more uniform culture which combines traits drawn from the traditions of each of the original communities.

Superdiversity A new high level of population diversity and structural complexity surpassing anything experienced in the past.

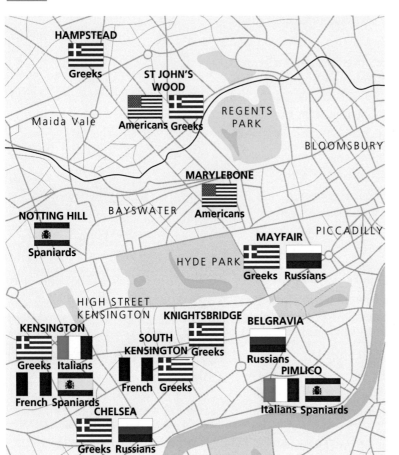

Connections and flows

Cultural changes to places in the UK can be viewed as a result of shifting flows of people in response to uneven global development and evolving global governance.

- International migration may often seem at first to have a simple economic cause, namely uneven development.
- Flows of people, however, are also a result of international political frameworks that allow free movement.

For localities in the UK that have undergone cultural changes on account of international migration, it is important to understand how these places are embedded in specific national and international contexts that have, during particular historical periods, permitted and actively fostered shifting flows of people.

▲ **Figure 3.22** Some places in London are home to a heterogeneous mix of different ethnic groups; but will differences lessen over time due to the melting pot effect?

🔑 KEY TERM

Post-accession migration The flow of economic migrants after a country has joined the EU.

In the immediate postwar decades, large numbers of post-colonial migrants arrived in the UK from British Empire (later Commonwealth) countries like India, Bangladesh and Uganda. Smaller numbers came from Nigeria, Kenya and other ex-British African territories (see Figure 3.23). More recently, **post-accession migration** from eastern Europe has introduced religious, linguistic and ethnic diversity to the UK, both in urban and rural areas.

- The first wave of migrants came to fill specific gaps in the labour force that opened up after the Second World War. Sometimes, migrants were recruited directly (London Underground held interviews for bus drivers in Kingston, Jamaica).
- There was still a large need for workers in heavy and light industry, especially in textile mills of the Midlands, Lancashire and Yorkshire. Labour demands were sometimes gendered – the Grunwick film processing factory in northwest London only recruited women of South Asian origin (the very poor working conditions they suffered led in 1976 to a protest involving 20,000 people).

- There were gaps in the skilled labour market too, notably within the ranks of the new National Health Service (UK doctors had not been trained in sufficient numbers during the 1930s and 1940s to fully staff the ambitious new NHS). Many doctors travelled to the UK from India, Pakistan and parts of Africa. One reason why it was so easy for the UK to achieve this aim was because of its language, customs and traditions. These had been introduced to British territories under colonial rule in the 1800s. Medical schools in India used the same textbooks as British teaching hospitals. The populations of ex-colonies spoke fluent English and showed an affinity with the British way of life. The UK was therefore able to take advantage of its past influence over these countries by advertising work opportunities to young Asians and Africans who were excited to move to the UK, following an education in schools where British history and culture would have been promoted.
- After the EU was enlarged in 2004, another wave of economic migration to the UK occurred. Large numbers of semi-skilled and skilled workers arrived from Poland for employment ranging from construction to dentistry. The extent to which these flows will now reverse on account of Brexit (see page 92) remains to be seen.

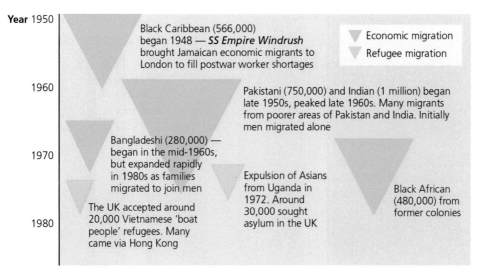

▲ **Figure 3.23** A timeline of post-colonial migration to the UK

Contrasting views on cultural change

Views about the desirability of migration and cultural change vary from place to place more than many people may realise. Deep schisms in British society were revealed by the UK referendum on EU membership. The country was not united in its desire to quit. Support for Brexit was high among pensioners, rural communities and urban areas in northern England, whereas younger voters, Scots and the cities of London and Cardiff favoured remaining.

One reason why communities in different places have varying views on recent post-accession migration is because of their own personal experience – or lack of it. Figure 3.24 provides a very interesting insight into how and why attitudes vary in the UK.

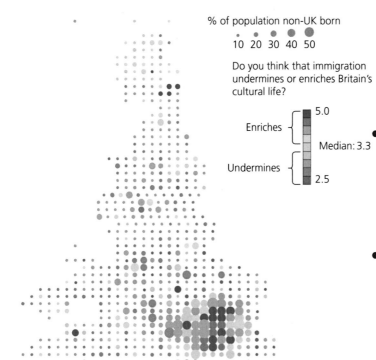

% of population non-UK born

· • ● ●
10 20 30 40 50

Do you think that immigration undermines or enriches Britain's cultural life?

Enriches
┌ 5.0
│
Median: 3.3

Undermines
└ 2.5

- Figure 3.24 shows public opinion at the level of UK parliamentary constituencies and data on the proportion of people in each constituency who are non-UK-born (a large circle means a high proportion).
- The colour of the dots shows people's responses in a survey to the question: 'Do you think that immigration undermines or enriches Britain's cultural life?' Green means more pro-migration and orange less.
- Researchers found a statistically significant positive correlation between the two variables: British-born people living in cities with a large migrant population, like London and Cardiff, tend to have more positive feelings (and were more likely to vote to remain in the EU in 2016). British-born people in places with fewer migrants tend to have more negative feelings.
- There are anomalies of course: Peterborough, Boston and Skegness have large migrant populations and relatively fewer people in these places apparently believe that migration enriches Britain's cultural life.

▲ **Figure 3.24** How attitudes towards immigration vary in relation to the size of local migrant populations (2011)

Evaluating the issue

▶ *Assessing the severity of spatial inequalities in the UK*

Identifying possible contexts, data sources and criteria for the assessment

Spatial inequalities can be explored using data collected at different geographic scales; a good starting point for this is research using books and articles by Danny Dorling, such as *Inequality and the 1% (2014).* A critical assessment of the severity of the UK's various spatial inequalities can be carried out in several logical steps by looking at:

1 large-scale inequalities between different regions and countries of the UK, including different parts of England, Scotland and Wales For many decades, geographers have been concerned with the existence of a so-called north–south divide (see page 79).
2 inequalities between different towns and cities, including close neighbours such as Stockport and Manchester
3 small-scale inequalities found within places, including disparities between different boroughs of the same city and smaller-scale differences between neighbouring postcodes or census output areas.

Inequalities manifest themselves in many different ways. For example, an assessment of inequality can explore:

- per capita wealth differences such as the value of people's houses or other assets
- per capita income measures (either estimated average earnings or GDP per capita of each region)
- unemployment (long-term or seasonal)
- other economic measures such as high-street closures or productivity per worker in different places
- social measures including average educational attainment, life expectancy and other health indicators.

Furthermore, spatial inequalities can sometimes appear more or less severe than they really are due to the selective way that quantitative and qualitative data are sometimes used. For example, one economic measure may show that a place is performing poorly, whereas an alternative measure might give a different picture. Biased reporting (quoting the 'good' data but omitting to mention the 'bad' data, or vice versa) is an all-too-common feature of social media platforms and some popular newspapers. Qualitative place representations in television shows and movies are often quite biased in the way they exaggerate the relative poverty or affluence of different places. This can make inequalities appear more severe than they really are.

Assessing inequalities *between* regions

The first, and largest, scale of enquiry is inequality between southeast England and other regions of the UK. The economic divide between the southeast and the rest of the country today is greater than at any time in the past. Historically, the value of houses has been widely used as a proxy for economic inequality. Variations between regions in the average cost of a home, along with the total value of the housing stock in each UK region, provide a crude but useful indicator of the economic health and wealth of each region.

- For much of the twentieth century, the cost of a London home was only about 25 per cent higher than the UK average. Since 2000, however, the divide has grown far wider. In

2017, the cost of an average London home was more than double the UK average value.

- In 2016, the total value of the housing stock in London and the southeast was £3 trillion. This was almost equivalent to half the value of all property in the rest of the UK. Moreover, inequality increased enormously in recent years: the aggregate value of London's homes doubled between 2012 and 2017, while homes in many northern cities including Burnley and Hartlepool fell in value. Large numbers of Londoners own housing assets worth £1 million or more: a fantastic 'windfall' for simply having lived there for many years!

- There is a great deal of further evidence for the north–south divide. Since the late 1970s, the southeast has constantly out-performed other regions, not only in terms of average earnings (£35,000 in the southeast in 2017 compared with national average of £27,000; more than half of all people earning £150,000 or more live in London and the southeast) but also health (life expectancy is 79 in the southeast; the 2017 UK average was 77). This persisting divide in part reflects how the southeast (excluding inner London) was less severely affected by deindustrialisation after the 1970s on account of its more diverse and resilient economy. Tertiary and quaternary industries helped keep employment losses below the national average. In contrast, those regions whose employment structure was traditionally dominated by manufacturing and mining suffered proportionately greater job losses on account of global shift and continue to score less well according to a range of socioeconomic measures.

When assessing the severity of the UK's large-scale regional inequalities, important things to recognise are (i) the persistence of the north–south divide and (ii) the way it has greatly strengthened in recent years.

Assessing inequalities *between* different towns and cities

Throughout the UK, strikingly divergent conditions can be found in settlements sited mere kilometres apart, such as Whitstable and Margate in eastern England. The former has a high-street vacancy rate of only 2 per cent, well below the national average. This historic town attracts tourists and is home to many high-income retired people ('affluent greys') who provide one-quarter of all spending. Some affluent Londoners have a second home in Whitstable. In contrast, Margate had a retail vacancy rate ten times higher in 2016. Parts of the town are in the bottom one per cent of England's most deprived areas and have an unemployment rate three times higher than neighbouring Whitstable.

This stark polarisation of fortunes among unequal neighbours is an emerging feature of present-day economic inequality in the UK. Why is this the case? One theory is explained below.

- The global financial crisis and the political austerity measures that followed (see page 86) have depressed earnings growth and created widespread economic 'drag' throughout the UK.
- However, some towns and cities have displayed far greater resilience against this challenge than others for a variety of local reasons.
- The result has been diverging growth patterns and the growth of inequality between resilient and less resilient settlements. Table 3.8 shows possible factors that may help this.

Factor	Examples
Reputation	Some tourism 'honeypot' sites, such as Brighton, Harrogate and Stratford-upon-Avon, remain popular with tourists, which helps to maintain local incomes.
Demography	Selective in-migration of 'affluent greys' or higher-salaried commuters has often focused on certain key settlements, such as Yorkshire's Hebden Bridge.
Public-sector activity	Settlements with a high proportion of public-sector workers, such as Sunderland, Birkenhead and Swansea, suffered above-average job losses when the UK Government began introducing austerity measures around 2010.
Private-sector activity	The most buoyant cities, such as Milton Keynes, Reading, Aberdeen and Bristol, continue to enjoy an abundance of private-sector employment and have lower vulnerability to public-sector job losses. In some cases, localised clusters of high-tech industries have been the secret for recent success (see page 137).
Accessibility	Highly accessible settlements, sited close to motorways, continue to receive inward investment from foreign companies. For example, both Bristol and Reading have above-average employment rates.

▲ **Table 3.8** Possible reasons why some settlements were more resilient than others following the global financial crisis

Assessing inequalities *within* local areas and neighbourhoods

There are marked inequalities within towns and cities too, not just between them. In the borough of Kensington and Chelsea, one of London's prime housing markets, the cost of a one-bedroom flat in certain places is ten times greater than in London's suburbs. Moreover, in 2017 the ever-growing average price of a Kensington house had reached 20 times an average UK citizen's salary (these extortionate prices have been driven by high demand from foreign-based investors seeking to purchase safe and durable assets).

Under the microscope, however, we find evidence for inequality in London *at an even more localised scale*. The social divide which exists in some of the capital's inner-urban areas has grown into a chasm. The Grenfell Tower fire of June 2017 – which led to the deaths of 71 people – drew attention to this disparity (see Figure 3.25).

- Built in 1974, Grenfell Tower was home to some of the remaining low-income individuals and families from London's Notting Hill and Ladbroke Grove neighbourhoods. These are places where a five-bedroom private home cost between £2 million and £20 million in 2018. Grenfell Tower residents could not afford these

prices; this is precisely why they had been housed there by Kensington and Chelsea Council.
- However, the poor management and hazardous design of Grenfell Tower resulted in disaster when a fire broke out because of a faulty fridge on one of the lower floors. Not enough had been done to mitigate fire hazards. There was only one communal staircase providing people with a means of escape and sprinklers had not been fitted.

▲ **Figure 3.25** Following the Grenfell Tower tragedy, the charred remains of the tower block (seen here from Latimer Road station) is viewed by many as a symbol of how unequal British society has become

Vulnerable elderly people who were housed on higher floors were unable to escape the blaze.

- Kensington and Chelsea Council had ignored warnings from residents in the years before the fire; critics see this as a symptom of wider *laissez-faire* management of inner-London places by local government which has allowed inequality to reach dizzying heights. The Grenfell Tower tragedy demonstrated all too vividly how Notting Hill is one of the most unequal places in Europe, if not the world.

Elsewhere in the UK, many settlements suffer from internal inequalities. In some cities, notably Manchester and Liverpool, the central business district (CBD) is relatively prosperous with many employment opportunities available. However, deprived neighbourhoods can still be found adjacent to the affluent city core. This is because communities in these deprived places often lack the skills needed to take advantage of new job opportunities in the CBD. They may also lack a car or find public transport unaffordable. This makes it harder for them to participate in the city economy.

Another way in which inequality manifests itself in urban areas is through changes affecting shopping streets in recent years. In Middleborough and Merthyr Tydfil, town centre shops have closed down in large numbers; demand has fallen so steeply that rents have fallen by about half since 2010. There are many reasons for this, including the growth of online retailing and a continuing trend for shoppers preferring out-of-town 'destination' retail parks such as the Trafford Centre near Manchester. Some town centres suffered badly from the failure of Woolworths and BHS (see page 96) and have experienced a domino effect of further shop closures due to the reduced pedestrian footfall following the loss of these 'anchor' stores.

Assessing the *reliability* of evidence for inequality

Chapter 2 explored the issue of bias in relation to place representations and it is important to return to this theme here. While assessing the severity of

inequality at varying scales, geographers must ascertain whether their data are valid and reliable. Media representations of affluent and poorer places often amplify and exaggerate; if you have ever seen the reality television show *Made in Chelsea* you will know that the lives of people living in places like Grenfell Tower do not form part of a narrative which is only concerned with the lives of affluent people in this part of London (Grenfell Tower is less than half a mile from where episodes of *Made in Chelsea* were filmed). In contrast, the portrayal of some towns and cities in the UK as problem places has sometimes exaggerated their poverty.

Representations of Liverpool are particularly interesting. Several competing narratives provide mixed messages about the severity of inequality between Liverpool and southern England. In 2012, Liverpool's inner-city district of Bootle was dubbed a 'terminal' place by the retail analyst Colliers International. A Centre for Cities report recently suggested that Liverpool as a whole is a 'vulnerable' city because of its over-reliance on public-sector employment. However, a competing survey by HSBC Commercial Banking characterises Liverpool as a potential 'supercity' which is poised to become a global hub on account of several important industrial growth clusters focused on green energy and finance (see page 144).

Clearly, it is difficult to assess the severity of inequalities when conflicting claims and datasets provide mixed messages about the extent of the economic prosperity or difficulty experienced by different settlements and neighbourhoods.

Arriving at an evidenced conclusion

In conclusion, spatial inequality is an enduring feature of British social and economic geography at different scales. These persisting inequalities have deepened in recent years. This is the result of the global financial crisis depressing growth in some places, especially many medium-sized northern towns and cities, while global investment flows have at the same time continued to inflate house prices disproportionately in selected highly prosperous places in southern England, such as London's Kensington and Chelsea.

The severity of spatial inequalities is often debatable due to conflicting evidence and claims, particularly when qualitative data are involved. What is beyond debate, however, is that the greatest inequalities in wealth, income and health are increasingly found *within* major cities and not *between* them. In 2017, the average price of a house in Manchester was £180,000 whereas the London figure was £470,000. However, the disparity between average Manchester and London prices pales in significance compared with house price inequality within London itself. The average price in top-ranking Kensington and Chelsea was almost £2 million (in the most desirable postcodes the figure is many times higher),

whereas bottom-ranking suburban Barking achieved just a fraction of this value (see Table 3.9). Finally, the Grenfell Tower tragedy was a damning indictment of the polarisation of wealth and rampant inequality between different postcodes in London and in many of the UK's other major cities too.

Rank	London borough	Average house price 2017
1	Kensington And Chelsea	£1.99 m
2	City of Westminster	£1.77 m
3	Camden	£1.07 m
31	Newham	£366,000
32	Bexley	£353,000
33	Barking and Dagenham	£298,000
	Average for all 33 boroughs	£470,000

▲ **Table 3.9** Average house prices in 2017 for the three most expensive and three least expensive of 33 London boroughs (Your Move data, March 2017)

🔑 KEY TERM

Central business district (CBD) The commercial centre of a settlement in which many shops and offices are located.

Chapter summary

✔ In recent decades, global connections and globalisation have driven structural change in cities and smaller-scale neighbourhoods throughout the UK and other developed countries.

✔ Manufacturing employment has dwindled in the UK (although output has sometimes remained strong due to automation); coal mining has all but vanished; service-sector employment has grown but is not immune to global shift and automation.

✔ Shifting flows of investment, resources and people have resulted in deindustrialisation. The accompanying cycle of deprivation, which may accelerate because of positive feedback, has brought severe social and environmental challenges to many inner-urban neighbourhoods.

✔ The GFC has combined with recently-arising technological and political developments, notably Brexit, to bring new challenges to less resilient places in the UK. More resilient places include fashionable districts in London and other large cities where house prices in some cases doubled after the GFC.

✔ Cultural and demographic changes have brought challenges and opportunities to different places; patterns of cultural diversity and the rising costs of an ageing population are issues of particular importance, both for national and local government.

✔ The UK has a very high level of social inequality which manifests itself at varying spatial scales, ranging from a broad north–south divide to the striking postcode disparities surrounding Grenfell Tower.

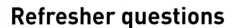

Refresher questions

1 What is meant by the following geographical terms? Global shift; deindustrialisation; automation.

2 Using examples, outline changes over time in employment and output for UK manufacturing and mining industries.

3 Using examples, outline the contribution originally made by particular industries to manufacturing employment in major UK cities.

4 Explain how the cycle of deprivation operates in deindustrialised places.

5 Outline the causes of the GFC and its consequences for one or more named places.

6 Explain why the Port Talbot steelworks in South Wales continues to face an uncertain future.

7 Using examples, explain how new technology creates challenges and opportunities for different places and their societies.

8 Outline how cultural diversity has changed in the UK in recent decades at both the national and local level.

9 Using examples, explain why the challenge of an ageing population is greater for some places in the UK than it is for other places.

10 Using examples, analyse the extent of inequality within and between different places in the UK.

Discussion activities

Your course may require you to use the place where you live as one of your course case studies. In pairs or groups, combine information from this chapter with your own local knowledge or research in order to discuss and make notes on the following themes.

1 What economic changes have occurred locally since the mid-1900s? You could draw and annotate a timeline to show changes in economic activity and people's employment.

2 What has caused these economic changes? What impact have globalisation and the GFC had on (i) your home place and (ii) the city or wider region it is embedded within?

3 How and why have the cultural diversity and age structure of your home place's population changed over time?

4 How does your home place compare economically and socially with other places, both (i) locally and (ii) nationally? Where would you position your home place on the spectrum of inequality found in the UK today?

5 Are you aware of any significant inward or outwards movement of investment and people linking where you live with other places, either nationally or globally? Do young people move away after leaving school, or is the place where you live somewhere that other people often migrate towards?

FIELDWORK FOCUS

Place changes, challenges and inequalities are a popular topic for independent investigations. These subjects lend themselves well to the collection of secondary data such as area census profiles and information gained using the Nomis service. There are numerous opportunities for primary data collection too, depending on the topic focus. Primary qualitative data, such as photographs showing environmental decay, can be used to support an analysis of deindustrialisation and the cycle of deprivation. Interviews with older people (oral accounts) can provide you with vital additional insights needed to create a profile of change in the place you are studying.

A *Investigating economic and social changes over time in your local neighbourhood.* A good starting point is www.nomisweb.co.uk/reports/lmp/ward2011/contents.aspx. Type in your postcode and away you go! If you're interested in investigating changes further back in time, it is possible to gain access to pre-1921 Census data at the household and street level. You can see who was living in your own house or street (if you live in an old house) more than a century ago. This is because the national census, which is conducted every ten years, has always obliged all households in the UK to provide details of the ages and occupations of each resident. The raw data are converted into aggregate statistics – this is what you can view at the Nomis website. After 100 years, however, the actual household questionnaires are made publicly available. Figure 3.26 and Table 3.10 show data extracted from the 1901 Census for a single home in the Yorkshire Village of Stanley. This could be the starting point for a study which compares people living in particular homes today with the families living there 100 years ago. Information taken from interviews with the current residents at those addresses could be contrasted with historical records from the 1901 or 1911 Censuses, available at www.freecen.org.uk and www.findmypast.co.uk.

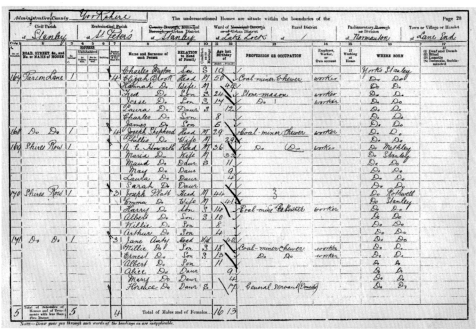

▲ **Figure 3.26** Extract from the 1901 Census record for a Yorkshire village

Name	Relation to head of family	Age	Occupation	Place of birth
Elijah Brook	Head	50	Coal miner	Stanley
Hannah Brook	Wife	49	-	Stanley
Fred Brook	Son	24	Stone mason	Stanley
Jesse Brook	Son	17	Stone mason	Stanley
Laura Brook	Daughter	12	-	Stanley
Charles Brook	Son	8	-	Stanley
James Brook	Son	5	-	Stanley

▲ **Table 3.10** The Brook family in 1901, as recorded by the census questionnaire shown in Figure 3.26

B *Patterns and causes of inequality for a place you are interested in studying.* Rather than exploring temporal changes, you may instead decide to focus on present-day spatial inequalities. The first question to ask yourself is whether you would like to investigate extremely localised patterns (comparing different streets in the same neighbourhood) or whether you would prefer to contrast two or more different parts of the same town or city. Census data will form an important part of your secondary research. Primary data collection possibilities include: interviews with estate agents; analysis of house price data (you may decide to stratify your sample in order to compare prices for different housing categories such as two-bedroom flats or four-bedroom houses); photographs taken by yourself which demonstrate contrasts in environmental quality. Zoopla is another possible data source.

Further reading

Brassed Off (1996) [film] Barnsley: Mark Herman.

Castells, M. (1996) *The Rise of the Network Society*. Oxford: Blackwell.

Centre for Cities (2018) *Cities Outlook 2018*. Available at: www.centreforcities.org/wp-content/uploads/2018/01/18-01-12-Final-Full-Cities-Outlook-2018.pdf [Accessed 21 February 2018].

Coe, N. and Jones, A. (2010) *The Economic Geography of the UK*. Sage: London.

Dorling, D. (2005) Hum*an Geography of the UK*. London: Sage.

The Full Monty (1997) [film] Sheffield: Peter Cattaneo.

Funderburg, L. (2013) The changing face of America. *National Geographic*.

Harvey, D. (1989) *The Condition of Postmodernity*. Oxford: Blackwell.

HSBC Business (2011) *The Future of Business 2011*. Available at: www.business.hsbc.co.uk/1/PA_esf-ca-app-content/content/pdfs/en/future_of_business_2011.pdf [Accessed 21 February 2018].

Massey, D (1984) *Spatial Divisions of Labour*. London: Macmillan.

Orwell, G. (1937, 2001) *The Road to Wigan Pier*. Harmondsworth: Penguin.

The place-remaking process

'Place remaking' is an umbrella term for various strategies to tackle negative economic changes and social inequalities affecting places. Using evidence and arguments related to a range of contexts, including Liverpool, Manchester and Hull city centres, this chapter:

- compares and contrasts different place-remaking approaches, including regeneration, reimaging and rebranding
- investigates how past legacies are used to support present-day plans for city centres and other places
- analyses different visions of contemporary place remaking, including 'smart places' and 'media places'
- assesses the importance of the role played by different stakeholders in the place-remaking process.

KEY CONCEPTS

Place remaking (or place making) This term describes all the collected physical, economic, social and cultural changes that can be carried out in a place, including redevelopment, reimaging and rebranding.

Actor network (or player network) A collaboration between different stakeholders (including local players and external agencies) who are seeking to collectively bring or resist change in a place or environment.

1 Place-remaking approaches, strategies and players

▶ *What place-remaking approaches and strategies can governments and other stakeholders adopt?*

The case for place remaking

Has any place been left unaffected by the sweeping global economic and technological changes and challenges of recent decades? It seems unlikely. Daily newspapers reported what seemed a rapid succession of industrial closures throughout the 1970s and 1980s (see Figure 4.1). The effects of these changes are clear: from the Clyde shipyards and cotton mills of Manchester to the coalfields of South Wales and London's Docklands, the majority of traditional jobs once found in these cities and regions are now long gone.

▲ **Figure 4.1** The 1970s were marked by factory closures, strikes and emergency measures like the 'three-day working week'

Moreover, some newer forms of work that supposedly 'replaced' traditional employment – such as jobs in data call centres and retail parks – do not necessarily provide local communities with a secure future. Telephone calls can be handled in the Philippines and India just as competently and for less money than in the UK. The arrival of Amazon and other online retailers has resulted in many UK shopping centres shedding workers (while increasing numbers of those who keep their jobs are asked to work on 'zero-hours' contracts).

The structural economic changes which have led to the shedding of secondary-sector and also some tertiary-sector employment in the UK are the result of broad historical processes of human development, globalisation and technological change. According to this view, both older and more recent waves of deindustrialisation were largely inevitable. But if we accept that change is unavoidable, does this mean politicians should step aside and let events take their course? Or must more be done to help communities wounded by deindustrialisation?

Do nothing or do *something*?

One view is that politicians should do nothing or as little as possible. This *laissez-faire* reasoning derives from a belief that economic change will ultimately leave society as a whole better off *even if it brings hardship in the short term*. New employers and forms of employment, so the argument goes, will eventually move to regions where there is a surplus of unemployed labour (and who are therefore prepared to work for lower wages than people in other more prosperous areas), thereby restoring economic growth in those places in the longer term. Alternatively, unemployed people can always 'get on a bike' (to adopt a notorious phrase used by one prominent 1980s Government minister) and seek work elsewhere.

While there is a logic to these arguments, they are not necessary supported by evidence.

- Many studies show that depressed areas stay depressed. A Sheffield Hallam University report from 2014 concluded that 30 years after many of the UK's mines were closed, coalfield communities continued to have the lowest employment rates in the UK.
- Deindustrialised areas have often suffered from environmental blight as a 'hangover' from the polluting industries that used to be there. Unfortunately, derelict warehouses and polluted waterways will often serve as a powerful deterrent to prospective investors in a place.
- Deindustrialised areas may suffer from a persisting skills shortage. The cycle of deprivation, as we have learned (see page 85), can lead to falling

standards in schools. New tertiary-sector and quaternary-sector employers may look elsewhere for a workforce unless there is some degree of government intervention.

- People are not always free to 'get on a bike' and look for work elsewhere. Many unemployed people in hard-hit deindustrialised places are also carers for elderly parents, some of whom suffer from long-term illnesses, including dementia. They are not at liberty to follow the logic of the marketplace and look for work elsewhere. Instead, they remain in their home place where they are desperately needed by others. Since 2010, UK Government austerity policies and welfare cuts have meant that many unemployed job-seekers face an increased burden of family care.

- Irrespective of what economic models and theories they believe in, wise politicians remain mindful of how unemployed voters who feel abandoned by the state can become a powerful political force. The Thatcher Government found itself in an often violent conflict with striking miners in 1984–85 (see page 82). Donald Trump's unexpected US presidential victory in 2016 can, in part, be attributed to the hardened political mindset of disillusioned unemployed voters in 'rust belt' states and cities (see Figure 4.2).

These are the realities of life for places adversely affected by exogenous economic and technical changes. As you might expect, most politicians – aside from those most thoroughly wedded to neoliberal ideology (page 75) – follow the general consensus that we must 'do something' to help manage changing places positively. But what? The remainder of this chapter explores varied aspects and examples of the place-remaking process, helping you form your own view about the strengths and weaknesses of different management decisions.

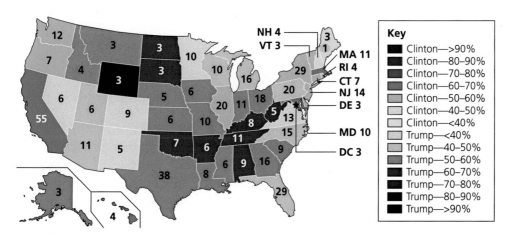

▲ **Figure 4.2** Donald Trump's victory over Hillary Clinton in the 2016 US presidential election was won by high levels of support in deindustrialised 'rust belt' states whose voters felt abandoned by mainstream politicians

The place-remaking process

The ongoing effort to remake British towns and cities has reshaped entirely the landscape and economy of many places. Often this has led to a profound change in place identity although, as we shall see, 'heritage' elements of past place identity often feature heavily in imagery used to advertise and promote present-day attractions.

The place-remaking process is at times complex and includes several distinctive elements (see Figure 4.3). It can be characterised in general terms as the *physical* and *psychological* reconstruction of a neighbourhood or larger-scale or urban area.

- The physical elements of remaking includes capital-intensive regeneration, particularly the construction of new flagship developments and infrastructure.
- The psychological elements consist of attempts to change the way people *perceive* a place and the *feelings* and *meanings* they attach to it (see Chapter 2). Tackling prejudice against places can be an uphill challenge though. This is why reimaging and rebranding strategies are so very important: by advertising and representing places in the media in appealing new ways, those responsible hope to 'win the hearts and minds' of potential visitors and investors.

🔑 KEY TERM

Flagship developments
These are usually high-profile and high-investment projects, such as new theatres, art galleries, museums, sports facilities or retail complexes. Iconic 'signature' buildings designed by well-known architects are intended to greatly enhance a city centre's reputation and image.

Remaking An umbrella term which brings together all the actions and interventions shown in this diagram. It includes capital-intensive attempts at regeneration and redevelopment, in addition to the reimaging and rebranding processes which allow places to be 'felt' and marketed as different or novel locations (while often retaining and protecting important 'heritage' features and meanings).

Regeneration and redevelopment Together, these include large-scale forms of state intervention and private sector inward investment which attempt to transform the fabric of place, often on a very large-scale. The aim is to attract new activity investment into an area while trying also to stimulate local enterprise. Often, there is a significant element of physical change to places, such as demolition, decontamination and land-use change (bringing new flagship retail and leisure facilities, housing and public transport systems). Ambitious schemes are concerned with affecting social change too, by providing new schools, universities or recreational spaces. *The strategic aim is environmental transformation.*

Rebranding A deliberate strategy usually adopted by formal-sector players which aims to reinvent places for the economic reason of marketing them (as tourist destinations, residential areas or sites for business investment). This may include a deliberate effort to shed an old or unwanted reputation. Local councils and tourist boards sometimes devote substantial sums of money to the creation of new advertising slogans or imagery, with paid assistance from a professional advertising or public relations company. Rebranding is often undertaken by a partnership of public-sector and private-sector investment, including major corporations. *The strategic aim is to increase 'sales'.*

Reimaging This process describes the cultural reinvention of places. It may involve formal-sector players creating new visual images of places in promotional materials. It also includes a wide range of ways in which places become informally represented as a result of the activities of artists, photographers, writers, musicians and film directors. Individually or collectively, they begin to change the image of a place in the texts they produce. In turn, this affects positively the perceptions of outside groups. Informal reimaging is at the heart of the gentrification process: in recent years, new residential incomers have been drawn to those inner-city neighbourhoods which the media represent as fashionable.
The strategic aim – if there is one – is akin to 'having a makeover'.

▲ **Figure 4.3** The main elements of place remaking (adapted from text by Alastair Owens).

Regeneration and redevelopment strategies

Extensive place remaking has been undertaken in the UK's largest urban areas such as London's Docklands (see Chapter 5, page 158) and Manchester city centre. In the case of Manchester, the need for redevelopment stemmed not only from global shift but also the IRA bombing of the city in 1996. A huge 1600 kg bomb was placed in a van and exploded outside the Arndale shopping centre, ripping through it with the force of an earthquake (a warning from the IRA meant the city centre could be evacuated and no-one was killed, though the BBC reported 200 people injured). But the bomb also left enormous spaces ready for redevelopment on a scale that would not normally be possible in the UK.

- The high cost of city-centre land ownership means that retail plots are typically small and it can be difficult to gain the consent of large numbers of property owners for redevelopment. In the case of Manchester, however, many acres of land were suddenly available for redevelopment following the bombing.
- A new plan for city-centre regeneration sought to restore some existing buildings while also introducing exciting architecture to the area, including a new bridge spanning Corporation Street. To the north of the city centre, Spinningfields was redeveloped as a so-called 'Canary Wharf of the north'.
- The first phase of regeneration took over six years to complete and in 2002 the final public open space to be completed, the refurbished Piccadilly Gardens, was opened.

▲ **Figure 4.4** The new Corporation Street bridge (left) was an important flagship development at the start of the regeneration process for Manchester, followed more recently by the development of the Spinningfields district (right)

Reaching an audience

Manchester quickly began to benefit from its fresh new image and reputation, helped in part by positive media portrayals of the regenerated city centre in popular 1990s television shows such as *Cold Feet*. Thanks to Manchester's successful bid for the 2002 Commonwealth Games, televised footage of the redeveloped city centre spread globally, reaching an audience of many more millions of viewers, including overseas students. Before long, young professionals and students had begun moving there in large numbers. The vice-chancellor of Manchester Metropolitan University explained in a 2016 interview: 'The city centre was dead in the 1990s. Now it is a 24/7 city. It feels exciting and feels on the up.'

Flagship developments

In the last two decades, more UK city centres have been extensively redeveloped by private-sector and public-sector stakeholders working in partnership. Major redevelopment schemes have introduced a mix of new land uses, including retail and office space, high-rise housing, entertainment and sports stadia, art galleries and recreational space. Bold and 'heroic' architectural statements (see Chapter 2's account of 'utopian' urban representations on page 60) have become an almost inevitable part of the mix.

- One of the newest eye-catching flagships is Birmingham's redesigned Grand Central station. Its completion in 2015 followed the redevelopment of the Bullring shopping centre (see Figure 4.5).
- Since 2004, the £106 million Wales Millennium Centre in Cardiff has celebrated Welsh culture in a modern setting, while also serving as an important focal point for the wider redevelopment and regeneration of Cardiff Bay.
- Newcastle's £70 million Gateshead Centre was completed in 2004 with National Lottery funding.
- London's 2012 Olympic venues played a pivotal part in the redevelopment of east London.

The flagship buildings described above and shown in Figures 4.4 and 4.5 have iconic designs that make striking use of modern architecture and materials. Although the sleek, angular shapes of some structures are similar in some ways to earlier brutalist designs (see page 32), this newer generation of modern

▼ **Figure 4.5** The futuristic Selfridges building is a flagship for Birmingham's redeveloped Bullring shopping centre. It was designed by architecture firm Future Systems and completed in 2003 at a cost of £60 million

buildings looks far brighter and more appealing (some are, in any cases, curved rather than angular). Major redevelopment works such as these often come with social costs attached, however. They obliterate pre-existing landscape features that may have been valued by the local community. Some people become displaced when homes are demolished and often too little is done to rehouse them in large-enough numbers in affordable housing units. These are themes which Chapter 5 explores further.

On completion, capital-intensive flagship developments, such as the Bullring or Gateshead Centre, serve as spaces for the consumption of goods, services or entertainment. In turn, they convey a powerful message nationally and internationally about the revived fortunes of the central urban areas they belong to.

- Images of these buildings often serve as representations of the cities they belong to in national and international media: a picture of the Wales Millennium Centre in a national newspaper might be described as 'a view of Cardiff', for instance (see also page 134).
- Central places are the most visible and frequently photographed places within any city and it is essential that city centres are redeveloped in ways which send the right 'post-industrial message' to potential visitors and investors.

Rebranding campaigns

In business, rebranding involves the use of press and media to raise product awareness through promotional literature and advertising campaigns. Usually, this accompanies a product reimaging 'makeover', the roll-out of a new slogan or logo and perhaps even a new product name. Increasingly, places are treated like products by both public-sector agencies (local councils and tourist boards) and businesses keen to see trade grow. One simple rule for products applies equally well to places: if worried about sales figures, try rebranding.

Rebranding reflects the urban entrepreneurialism that lies at the heart of place remaking in contemporary capitalist societies.

- Competing societies are all trying to grab a share of footloose global capital to invest in their locality in order to revitalise it (see Figure 4.6).
- Rebranding therefore becomes seen as essential for maintaining visibility within a global economy where myriad places are vying for attention in what some view as a 'zero-sum' game (meaning that if one place attracts a larger share of tourist investment flows, other places may get less accordingly).
- If a place can successfully use rebranding to make itself better known to external clients and stakeholders, it becomes a more powerful locality with an ability to 'act at a distance' within a global systems framework.

 KEY TERM

Urban entrepreneurialism Associated with the writing of David Harvey, this term describes a model of urban governance which promotes economic growth by encouraging the private sector to fund new urban housing or industrial developments in return for profit (rather than expecting public-sector funding to pay for everything).

Rebranding Liverpool as a European 'Capital of Culture'

Rebranding was an essential element of plans for the economic rejuvenation of Liverpool's city centre. The redevelopment of the Albert Dock for tourism in the mid-1980s had given the city its first real glimpse of what a post-industrial urban economy might look like. By the 1990s, a broader vision of Liverpool rebranded as a city of culture was gaining support from growing numbers of local stakeholders. According to this view, the *psychological* rebranding of the city as a cultural heartland – rather than a traditional industrial hub – was going to be just as important as the *physical* redevelopment of the city centre.

Rebranding of Liverpool city centre was achieved in two steps.

1 The city council nominated Liverpool for two important awards and won both. The city centre gained UNESCO World Heritage Site status in 2004 (see also page 69). Also, the coveted title of 'European Capital of Culture' was granted to Liverpool by the European Union for the year 2008. In both cases, it is important to note how the international character of these designations greatly enhanced Liverpool's global reputation and ability to attract international visitors.

2 An organisation called Liverpool Culture Company was created by Liverpool City Council and entrusted with responsibility for the development of cultural events in the run-up to 2008. Liverpool Culture Company's remit to 'deliver the culture

The Hebrides: "Experience life on the edge"

Moray: "Malt whisky country"

Scotland: "A spirit of its own"

Clackmannanshire: "The wee county"

Fife: "Kingdom of life"

Glasgow: "People make Glasgow"

Arran: "Island time in no time"

Edinburgh: "Incrediburgh"

Kirkcudbright: "Scenic fishing and an artistic heritage"

Dumfries & Galloway: "The natural place"

Northumberland: "England's most tranquil county"

Cumbria & the Lake District: "Make time for you"

Yorkshire: "Alive with opportunity"

Southport: "Day time, night time, great time"

Leeds: "Live it, love it"

Liverpool: "City of culture"

Manchester: "The original modern city"

Peterborough: "A city to surprise and delight"

Mid-Wales: "Because Mid-Wales is as unique as you are"

Birmingham: "The global city with a local heart"

Norwich: "A fine city"

Norfolk: "Time to explore"

Cardiff: "Gateway to Wales"

Bedfordshire: "We ♥ Bedfordshire"

Essex: "Explore, experience and enjoy"

London "Totally LondON"

Wiltshire: "A rich history"

Plymouth: "Positively Plymouth"

Cornwall: "Cornwall for ever"

Poole: "Surf, rest and play"

Eastbourne: "The sunshine coast"

▲ **Figure 4.6** A snapshot of competing branding slogans used at various times since 2010 by different cities and regions of the UK in a perceived zero-sum competition for investment and visitors

programme up to and beyond 2008' was achieved through a series of advertisements designed to change and challenge perceptions of Liverpool, both in the UK and globally (see Figure 4.7). One press release read: 'The Liverpool 2008 opportunity will build on our universal reputation and create a world-class brand based on dynamic creativity.' Liverpool Culture Company made sure that Capital of Culture features appeared in a string of high-profile media publications around the world, including *India Weekly*, *Le Temps*, the *Washington Post* and *Time*.

▲ **Figure 4.7** The Capital of Culture initiative is credited with providing Liverpool city centre with a new sense of place that is rooted in cultural heritage, such as 'The Three Graces' skyline shown here

The reimaging process

As the example of Manchester used on pages 121–122 shows, there is more to place remaking than structural works. Manchester's successful reinvention of itself, as we have learned, was also aided by the way its city centre was portrayed in the media during the Commonwealth Games and as a backdrop for popular television shows. The city's enduring role as a home for music subcultures was helpful in this respect too. The Hacienda Club (see page 54) endured well into the 1990s; Manchester was also portrayed by entertainment media as the home place for Britpop, a 1990s musical movement, thanks in large part to the national popularity of local band Oasis. The great success of two outstanding football clubs raised Manchester's media profile higher still.

Indeed, by some measures of economic growth Manchester has been the UK's most prosperous city in recent years after London. Its economy has grown faster than the UK as a whole and almost doubled in size over the past 20 years; the city's population has grown by one-quarter too.

It is important to recognise, however, that perceptions of Manchester as a vibrant and successful city in the late 1990s and early 2000s arguably relied as much on informal reimaging (achieved through the popularity of Oasis and Manchester United's footballers) as it did on the formal regeneration and redevelopment processes that had begun after the IRA bombing. Reimaging is not a process that can always be controlled and managed formally though. Indeed, formal-sector players sometimes struggled to control imagery they disapproved of – not all of the associations between Britpop for Manchester were positive, for instance, and the movement was widely criticised for its 'laddish' culture.

Despite this caveat, the overall success of Manchester's reimaging is evident in the way more than half of the city's university students originate from distant parts of the UK or overseas. They are drawn not only by the quality of teaching but also – as students tend to be – by positive media-driven perceptions of the city and high expectations of what life in Manchester will feel like. Moreover, seven-in-ten university students stay in Greater Manchester after graduation. This is one reason why Manchester generated £31,000 of economic growth per person in 2017, double that of neighbouring Rochdale and Oldham. By these and other measures, place-remaking processes have had very positive outcomes for central Manchester.

CONTEMPORARY CASE STUDY: THE UK CITY OF CULTURE AWARD

'UK City of Culture' is an official title awarded once every four years to a city in the UK for a period of one year. In 2017, the designation was given to Hull. This award does not bring any direct government funding (although Hull did receive a £3 million Heritage Lottery Fund grant) but is nonetheless seen as an excellent way of improving any city's image, in turn encouraging new private-sector investment from local, national and international sources and stakeholders. Like the European Capital of Culture prize that it is modelled on and which Liverpool won (see page 124), the UK City of Culture accolade is first and foremost a stimulus for the regeneration of deindustrialised cities through culture and the arts.

Hull's year as UK City of Culture

In the run-up to 2017, Hull City Council put culture at the heart of their regeneration plan, making it instrumental in their efforts to transform the city.

▲ **Figure 4.8** Daily artistic or cultural events were staged in Hull throughout 2017, the city's year as 'City of Culture' title-holder

■ In an interview with one national newspaper, the leader of Hull City Council said: 'We set out as a city to say to visitors "have a good closer look, this is actually a very nice place to come to" – and we changed people's views.' This clearly demonstrates the importance of place perceptions as a focus for action among the key players responsible for place remaking.

■ By one estimate, Hull's cultural programme of 365 days of art, theatre and music in 2017 brought a £60 million boost to its economy (see Figure 4.8); another estimate suggests that the city received £1 billion of new investment in the run-up to 2017 (the award of the title was announced four years in advance in 2013).

Competing for the 2021 UK City of Culture award

The final contenders for the 2021 title were Stoke-on-Trent, Swansea, Sunderland, Coventry and Paisley. All hoped to win the title in order attract investment and stimulate entrepreneurial activity among local community groups. Each of these deindustrialised cities had a reason to want the award badly.

■ As one newspaper article put it: 'If there was ever a town with image problems it is Paisley, the former textile powerhouse outside Glasgow that for decades has been a byword for drugs, deprivation and high street crime. Until recently, many of the most talented alumni of the celebrated youth theatre were reluctant to admit their Paisley roots. The town was once one of Scotland's richest but was devastated by the closure of textile mills in the later 1900s.'

■ In the event, Coventry was named UK City of Culture 2021. Coventry is the birthplace of poet Philip Larkin and The Specials, whose negative musical portrayal of Coventry – *Ghost Town* – was discussed on page 53. It will be interesting to see how beneficial this new – and far more positive – media portrayal will turn out to be for the city.

Stakeholders in the place-remaking process

An important part of place-remaking studies involves the analysis and evaluation of different stakeholders' input into the process. Stakeholders – alternatively described as 'players' or 'actors' – can be defined as all of the individuals, groups and organisations with involvement or interest in a particular place or issue. Place remaking is an often expensive business that usually involves a wide range of stakeholders working collaboratively.

Some players provide capital; others may contribute marketing skills or other areas of professional expertise. Typically, a large-scale place-remaking initiative will be a public–private partnership, meaning that there is an input of capital and expertise from both profit-seeking businesses and public-sector players (government). Public-sector involvement in the largest regeneration projects is typically **multi-scalar** (see Figure 4.9).

 **KEY TERM**

Multi-scalar When referring to governance, this means that a range of local, regional, national and even international (EU or UN) government actors and agencies are involved.

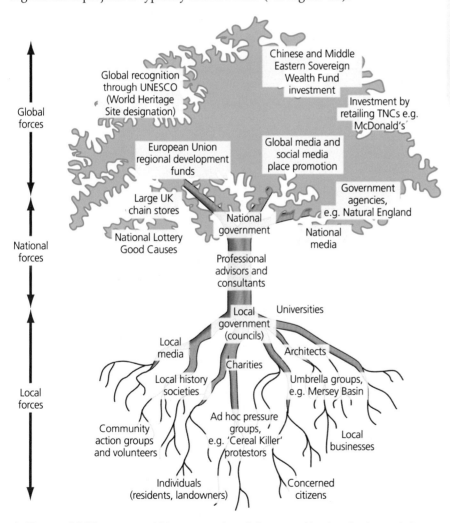

▲ **Figure 4.9** The most ambitious examples of place remaking involve interrelations between many different players at varying scales of action

Government interest in place remaking usually extends well beyond matters of money and additionally encompasses social and environmental goals. Parallel objectives alongside the aim of bringing new employment to an area typically include:

- an increased supply of affordable housing and improved provision of education or skills for local people
- environmental goals such as improved water quality in previously industrialised places.

Global and international stakeholders

Larger-scale place remaking usually has international dimensions: transnational corporations, sovereign wealth funds and diaspora communities can all be important global stakeholders in the place-remaking process (see Table 4.1). Their involvement demonstrates once more how local places are affected by shifting connections with external forces across often considerable distances.

Stakeholder	Role in place remaking
Transnational corporations (TNCs) and retail developments	■ Global brands now outnumber UK retailers on London's Regent Street. Large global brands such as Apple, Gap, Superdry, Adidas and Levi's are an important presence too in new retail developments such as Westfield (London) and St David's (Cardiff). ■ These companies rely on pedestrian footfall and will pay a premium for a site with high visibility; TNC chain stores therefore play a vital role in the financial fortunes of new shopping centres. ■ The Westfield Corporation itself is a foreign investor in the UK (and other countries); its headquarters are in Australia (although a deal was announced in 2017 to sell Westfield to France's Unibail-Rodamco).
Sovereign wealth fund (SWF) investment in place remaking	■ Some state governments buy foreign assets using their SWFs. These are global-scale 'piggy banks' used by these countries to help build global influence and diversify their income sources. Only a minority of countries operate SWFs and they are often countries with oil and gas revenues, such as Norway and Qatar. ■ SWFs own a range of assets in regenerated British cities including infrastructure, prime real estate and football teams (see Figure 4.10); Qatar's SWF owns a chunk of Canary Wharf; a 25 per cent stake in Regent Street is owned by the Norwegian SWF; China is expected to be an important backer for the High Speed 2 railway (see page 53).
Diaspora involvement in place remaking	■ Another international dimension to place remaking is the role sometimes played by diasporas (the worldwide scattering or dispersal of a particular nation's migrant population and their descendants). ■ The UK's 'Celtic fringes' (Scotland, Wales and Northern Ireland) have all produced significant global diasporas despite their relatively small population sizes. Place remaking in cities such as Dublin, Cardiff and Glasgow may therefore involve reaching out globally to people of Irish, Welsh and Scottish descent respectively (in order to find investors and prospective victors).

▲ **Table 4.1** International forces can have an important role to play in the place-remaking process

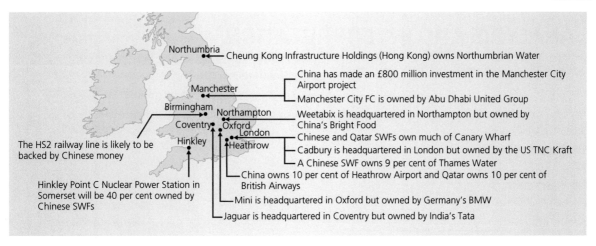

Northumbria — Cheung Kong Infrastructure Holdings (Hong Kong) owns Northumbrian Water

Manchester
China has made an £800 million investment in the Manchester City Airport project
Manchester City FC is owned by Abu Dhabi United Group

Birmingham
Northampton — Weetabix is headquartered in Northampton but owned by China's Bright Food

Coventry
Oxford
London
Heathrow

The HS2 railway line is likely to be backed by Chinese money

Hinkley

Chinese and Qatar SWFs own much of Canary Wharf
Cadbury is headquartered in London but owned by the US TNC Kraft
A Chinese SWF owns 9 per cent of Thames Water
China owns 10 per cent of Heathrow Airport and Qatar owns 10 per cent of British Airways
Mini is headquartered in Oxford but owned by Germany's BMW
Jaguar is headquartered in Coventry but owned by India's Tata

Hinkley Point C Nuclear Power Station in Somerset will be 40 per cent owned by Chinese SWFs

▲ **Figure 4.10** Sovereign wealth funds are global institutions that play an increasingly important role in place remaking throughout the UK

CONTEMPORARY CASE STUDY: THE IMPORTANCE OF THE SCOTTISH DIASPORA FOR PLACE REMAKING IN SCOTLAND

Scotland is home to just 4.5 million people, yet up to 40 million individuals living worldwide claim Scottish ancestry, including people in the USA, Canada, New Zealand, Argentina and Brazil.

Worldwide dispersal of Scottish people dates back to the late eighteenth and nineteenth centuries, when Scotland's rural poor suffered an internal and subsequently international displacement called the Highland Clearances.

■ As a result of changes in land ownership (or what might be described today as a 'land grab'), 150,000 Scots were evicted from their crofting homes and made landless between 1780 and 1880. Many sought improved prospects overseas.

■ At this time, the UK Government was offering British citizens free passage to the New World – to Canada, Australia and New Zealand – partly to ease population pressure at home and partly to shore up Britain's global territories.

■ Some richer Scottish families also migrated overseas in the early industrial period – many large slave plantations in Jamaica had Scots owners.

These past movements explain the present-day existence of a vast Scottish diaspora that can be enormously beneficial for places in Scotland. Online ancestry websites enable people living all over the world to trace their roots back to Scotland; many become keen to visit. This is an enormous prize for the Scottish tourist industry: 2.75 million international visitors spent £1.9 billion in 2016.

■ Tourist-orientated place-remaking strategies in Scotland often play up to the expectations of visitors with Scottish roots who want to see 'tartan and bagpipes'. However, some young people living in Scotland object to this so-called 'fetishisation' of the past; they fear their nation risks becoming culturally fossilised and would prefer to be represented as a forward-looking independent European nation. Figure 4.11 offers a glimpse of these contested place representations.

■ GlobalScot is a website run by government-funded Scottish Enterprise. This is an alternative approach to global networking which relies less heavily on the cultural reconstruction of heritage. Instead, members of the Scottish diaspora are encouraged to trade, deal, invest and forge business partnerships with one another. The GlobalScot website largely eschews 'Tartanry' in favour of contemporary financial and technological imagery.

▲ **Figure 4.11** A display of 'Tartanry' – heritage products whose intended market includes the global Scots diaspora

ANALYSIS AND INTERPRETATION

Study Figure 4.12, which shows the total internal migration between UK cities of 30–39-year-olds in 2015.

(a) Estimate the number of people who arrived in Birmingham in 2015.

GUIDANCE

You can use the data for Manchester and London to help you estimate an approximate value.

(b) Using evidence from Figure 4.12, assess the view that London benefits more than other UK cities from internal migration.

GUIDANCE

The instruction 'Using evidence from Figure 4.12' sends a signal that this question should be answered by applying analytical skills to the resource instead of using recalled case study knowledge. The data can be analysed in ways which allow a case to be made supporting the view that London has benefited most. Most important of all, London has received more migrants than any other city. However, the data can be manipulated in alternative ways that help us reject the proposition. First, we can make estimates of the net gain or loss each city has experienced. London has actually made a very small net loss whereas both Birmingham and Manchester have made net gains. Birmingham in particular has done very well. Second, London is seven times larger than any of the other cities shown and this could be taken into account when judging success.

(c) Suggest how recent rebranding strategies may have contributed to the movements towards northern cities shown in Figure 4.12.

GUIDANCE

One way to approach answering this question is to start by explaining what rebranding strategies hope to achieve. The aim is to change the perception that outsiders have of a local place or larger-scale city. In particular, presenting somewhere as being a particularly vibrant and rejuvenated destination can prove attractive to migrants in other places. A good point to suggest would be that city councils and other stakeholders in some northern cities have perhaps advertised and marketed these destinations in positive ways, which has affected the perceptions of Londoners thereby prompting them to relocate. Also, the 30–39 age bracket of the migrants shown in Figure 4.12 is important to think about as part of your answer. These are people at a stage in their life cycle where they may be thinking of starting a family or already have young children. Rebranding strategies could have been targeted at this particular demographic group. What kinds of positive media images or messages might help convince a 30–39-year-old cohort of Londoners to move north?

Internal migration of 30-somethings
Total moves between cities of 30–39-year-olds (year ending June 2015)

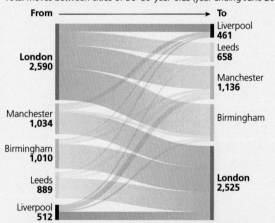

FT graphic   Sources: DCLG, ONS, HM Land Registry, Colliers

▲ **Figure 4.12** Age-selective (30–39) migration flows linking major UK cities, 2015

Cultural heritage and place-remaking processes

▶ *How can present-day growth be supported using local history and legacies of past economic development?*

Place remaking that draws on the past

Different local players and external agencies – ranging from government agencies and businesses to local community groups – will frequently draw on established 'heritage' place meanings when carrying out redevelopment and rebranding work. These are the important legacies of past development processes which have created persisting place meanings for people. Chapter 1 explored how football teams that today form an iconic part of the cultural landscapes of major cities are often a legacy of past industrial development. Sheffield's metals and steel-working history remains clearly visible in the badge of one of its teams (see Figure 1.6, page 12); the Liver bird emblem which appears on Liverpool Football Club's team badge has strong traditional associations with the city's maritime trading history.

In these and countless other ways, the past plays a vital role in the ongoing (re)construction of contemporary identities. Many local places with a long and rich history, like the larger-scale regions and nations they are embedded within, celebrate their heritage and 'origin stories' in ways that foster present-day social identity and community cohesion, much as Benedict Anderson argued (see page 45). Heritage initiatives contribute to the place-remaking process in different locations using strategies which, as we have glimpsed in previous chapters, span:

- industrial heritage (including engineering feats, machinery, canals and railways)
- creative arts (local associations with famous writers, poets and painters)
- architecture (and iconic historical buildings)
- regional food and drink legacies (such as the peat-smoked food and alcohol that helps support the tourist industry in Scotland's Western Isles, outlined previously on page 44).

The visitor experience

Truly effective rebranding using heritage does more than merely embrace nostalgia. Successful attempts to tap into local history will make use of past place legacies while at the same time updating the visitor experience by blending present-day technologies, trends and fashions. For instance, the success of Terry Deary's *Horrible Histories* franchise has prompted the management of many museums and heritage attractions to review their own histories, archives and resources in order to find particularly 'horrible' (that is, unpleasant or gory) exhibits and stories to use as 'headline' advertising in their promotional materials.

- The London Dungeon exhibition became very popular in the 1990s after capitalising on the local area's macabre associations with Jack the Ripper and Sweeney Todd.
- Manchester's Imperial War Museum (North) attempted to bring alive the city's wartime history by staging a 'Horrible Histories: Blitzed Brits' exhibition in 2015, while the National Maritime Museum in Greenwich has many exhibits related to death and violence in wartime, such as cat-o-nine tails and surgical saws.
- Permanent displays at the International Slavery Museum in Liverpool and in Bristol's M Shed museum show us 'horrible histories' of a very different kind. Both cities grew rich at the expense of the transatlantic slave trade. There are campaigns in Bristol to change the names of numerous landmarks such as Colston Hall, which is named after a slave trader (see Figure 4.13). Critics say that too often industrial heritage has been sanitised and there is not enough attention paid, even today, to the unethical sources of wealth that funded the British 'golden age of prosperity' during the 1700s and 1800s.

▲ **Figure 4.13** Colston Hall in Bristol is named after Edward Colston, a slave trader. Many people believe the name should be changed; others say it would be wrong to 'airbrush' over historical place connections, even if the truth hurts

Some flagship developments in large cities, including those described on page 122, play an important role as venues where heritage events can be staged. Despite their futuristic exterior cladding, much of what happens inside these spaces is really a celebration of the past, including music and theatre. Wales Millennium Centre in Cardiff in particular was envisaged as a cultural hub where Welsh national identity could be celebrated (Figure 4.15). Table 4.2 provides further examples of contemporary redevelopment and regeneration that has made full use of local heritage and historical legacies at varying spatial scales.

Location	Redevelopment/regeneration efforts
Central London	■ Chapter 2 (see page 30) examined some of the issues surrounding the redevelopment of central London necessitated by the construction of Crossrail. New investment has brought state-of-the-art improvements to railway stations along the route. In particular, the redesigned Tottenham Court Road station is now a catalyst for neighbourhood redevelopment and modern reimaging. ■ But the construction of Crossrail has also created an opportunity to celebrate *past* development processes in the city of London. Many archaeological artefacts have been retrieved from seven years of excavation work; thousands of objects spanning 8000 years have been collected in one of London's newer visitor attractions, the Museum of London Docklands. The displays include decapitated Roman-era skeletons and evidence of other 'horrible histories'.
Hastings	■ The Royal Institute of British Architects awarded its 2016 annual prize to the redeveloped seaside pier at Hastings. This town on the south coast of England has struggled for many years with a poor reputation and high levels of child poverty. The rebuilt and reimaged pier will, it is hoped, help change how Hastings is perceived by outsiders. ■ Originally constructed in 1872, Hastings' Victorian pier is an important heritage feature that has been thoroughly updated using the latest building materials (the new pier has a timber and glass structure with an innovative angular design). There is therefore both continuity and change in Hastings' new tourist image.
Plymouth	■ Plymouth played a key role in British naval history, particularly during the Elizabethan war with Spain many centuries ago. Unsurprisingly, maritime history also has enduring importance for the city's economy. For example, the Royal William Yard in Plymouth has been redeveloped as a leisure and residential site by an organisation called Urban Splash (see Figure 4.14). New bars and flats occupy the same space where the British Navy's supplies were once stored. ■ A rebranding campaign called 'Positively Plymouth' was briefly used with the hope of attracting more visitors to Royal William Yard and other places in Plymouth. ■ However, in a BBC Radio 4 series about changing places, the writer Will Self observed that the reimaging of Plymouth had not always benefited its original inhabitants. People who were 'decanted' elsewhere when areas like Devonport and Royal William Yard were being redeveloped did not always return afterwards.
East Devon and Dorset Coast	■ This final example of heritage-based rebranding is worthy of study because of its larger-scale spatial and temporal dimensions. A 200 km stretch of southwest England was rebranded as 'The Jurassic Coast' in 2001 after winning UNESCO World Heritage Site status because of its geology and fossils. ■ In this instance, present-day development plans are drawing on a geological history that spans hundreds of millions of years.

▲ **Table 4.2** Examples of how the past is an important influence on present-day place remaking

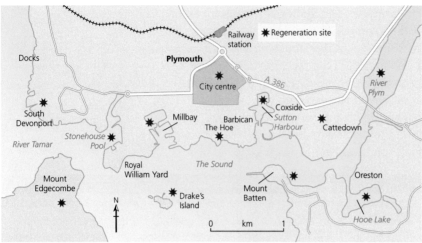

▲ **Figure 4.16** Built in 1848, Birmingham's Broad Street Presbyterian Church closed its doors to churchgoers in the twentieth century and reopened them to Ministry of Sound nightclubbers before more recently becoming Popworld.

▲ **Figure 4.14** Plymouth's naval history provided the 'raw material' for an ambitious rebranding and reimaging campaign spanning the entire seafront

▲ **Figure 4.15** Wales Millennium Centre and Cardiff Bay

Don't judge a book by its cover

Redevelopment schemes do not always result in old buildings being replaced with new ones. Instead, historic buildings can be retained with a change in their land use. In fact, the façades of old properties may be highly valued precisely because of the feeling of authenticity they give to places where heritage and culture are supposed to be major selling points. Old buildings may even be protected under planning law if they have been placed on the Statutory List of Buildings of Special Architectural or Historic Interest (see also page 33).

But these properties can sometimes be given a brand new function without changing their outward appearance significantly. Many old factories and warehouses – such as those at Plymouth's Royal William Yard – have gained a new lease of life as shops, restaurants or flats. The Broad Street Presbyterian Church in Birmingham has been converted into a nightclub despite being a Grade II listed building (see Figure 4.16). Often, there is both continuity and change in the way places are remade.

Heritage manufacturing

Place remaking does not depend exclusively on the promotion of tertiary-sector leisure activities. Despite the fact that the UK is, in general, characterised as a deindustrialised country, it still retains a small thriving manufacturing sector, as Chapter 3 explained (see page 81). In common with heritage tourism, a blending of *past and present* economic development processes can frequently be seen in the way some UK manufacturing industries have adapted to global trading conditions in the twenty-first century.

Post-Fordist manufacturing

For the UK, global shift has meant the virtual disappearance of assembly-line or 'Fordist' manufacturing activity – a name gained from the early-twentieth-century car-making factories of Henry Ford (see Table 4.3). However, manufacturing work has remained viable, in some cases growing from strength to strength, by adopting an alternative approach to production called post-Fordist activity (see page 25).

Fordist (fixed assembly line)	Post-Fordist (flexible production)
Mass production of a single product	Small batch production
Product design rarely changes	Frequent changes to design
Large stocks and warehousing	Just-in-time (JIT) deliveries of parts and goods
Single task performed by worker	Workers can multi-task

▲ **Table 4.3** Fordist and post-Fordist activities

Essentially, post-Fordist manufacturing firms produce small batches of varied or exclusive products. Examples include luxury and specialist clothing and footwear, including outdoor jackets and climbing boots. Around 100,000 people work in the UK's premium fashion and textiles industries. The luxury clothes, furnishing, rugs and carpets they make contribute £7 billion to the UK economy annually, according to one estimate. Table 4.4 shows a handful of examples. In many cases, present-day activities are clearly related to work carried out in these same places during earlier centuries.

Some of the activities in Table 4.4 could also be described as artisan work. Historically, the artisan was a skilled worker who made things by hand, often to very high quality. Blacksmiths, metal-workers, jewellery-makers and the most skilled weavers were all artisans. During the 1800s and early 1900s, industrialisation, mass production and assembly lines came to dominate British industry and artisan production often ceased.

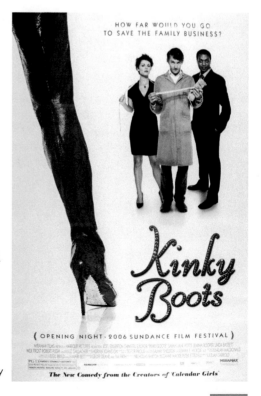

▶ **Figure 4.17** The 'Kinky Boots' story celebrates the remaking of a Northamptonshire town's economy in a quite unexpected artisanal way

Place	Post-Fordist activities
Guiseley, Yorkshire	Premium-quality cloth is hand-woven in the old Abraham Moon mill in the village of Guiseley near Leeds. TNCs Hugo Boss and Prada have both used this cloth in their luxury products.
Earls Barton, Northamptonshire	WJ Brookes Ltd in Earls Barton mass-produced men's shoes for 110 years. Recently, the company abandoned its assembly-line products. Workers were re-trained to hand-make expensive high-heeled boots for men. The film and stage show *Kinky Boots* celebrates this real-life story (see Figure 4.17).
Wickhamford, Worcestershire	Douk is a small company in Wickhamford, Worcestershire that has made hand-built snowboards since 2009. Prices start at round £400 per board – clearly, this is a very different manufacturing business model from, say, mass-produced Chinese-made skateboards sold on Amazon for £15. Snowboarding's unique styles, allied with its target audience's relatively high purchasing power, makes it a perfect fit with the 'logic' of post-Fordist capitalism.

▲ **Table 4.4** Places where post-Fordist 'heritage' manufacturing industries have thrived despite the challenge of global shift

- It is interesting to reflect on the way things have, to some extent, turned full circle: past development processes have been reinvented for present-day purposes.
- Skilful craftspeople, sometimes manufacturing items by hand, are thriving again throughout contemporary Britain. Later in the book, Chapter 6 investigates examples of contemporary rural industries that have revived traditional artisanal crafts and food making.

Remaking contemporary places

▶ *What role can contemporary design and technology play in place remaking?*

Out with the old, in with the new

Place remaking is not always to do with the past. There are plenty of interesting cases where redevelopment and reimaging efforts are far more focused on contemporary design and innovation.

- Not all places have much history to celebrate in any case: think of the new towns built from scratch in the UK during previous decades – Milton Keynes (see Figure 1.4, page 9) only came into being in 1967.
- Remember too that countries have, on occasion, conjured entirely new capital cities out of nothing, including Brasília (Brazil) and Abuja (Nigeria), thereby helping them project a modern, 'open for business' image.

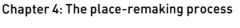

- For these settlements and smaller-scale local places (such as new housing estates or industrial parks), their sense of identity stems from their newness and *lack* of historical 'baggage'. Well-planned and carefully-designed places can visibly incorporate the latest new technologies into their very fabric.
- In Chapter 2, we encountered the spirit of forward-looking 'utopian urbanism' that embraces bold, modern design 'statements' (see page 59).

It is not only new places that can benefit from innovative design, planning and technology, however. These qualities can be incorporated into older settlements too through the process of **place retrofitting** – a phrase which describes alterations made to buildings and infrastructure on a sufficiently large scale for new technologies to become an important characteristic of an entire neighbourhood or larger area.

There are several ways in which contemporary research, design and technology can contribute to the construction of new places or the retrofitting of older places. This section explores the construction of three examples of contemporary design and planning:

- *smart places* – where fast broadband is being used to deliver intelligently designed city services and 'the internet of things', potentially allowing places to become quaternary industry or **knowledge economy** hubs
- *ecocities and eco-districts* – where technology is being used to minimise environmental 'footprints', for example by maximising energy efficiency or enhancing recycling
- *media places* – where landscapes that have appeared in recent or classic films and television shows become magnets for fans.

Smart places

Some cities now promote themselves as 'technology hubs' where digital or creative industries can cluster. In a few cases this involves rebranding as a 'smart city' where the latest technology is used to run city services more intelligently.

- The UK is home to impressive examples of clusters formed of life sciences industries, including areas in Cambridge, Edinburgh and Leeds. The Cambridge cluster alone is home to more than 400 life sciences companies. Their growth is attributable to an *ad hoc* mixture of planning and local university support and scientific excellence. The UK Government has also played an important role supporting these industries, most recently with its Life Sciences Industrial Strategy.
- A recent £75 million partnership between the University of Bristol and the city council has brought exceptionally fast broadband to Bristol. A cluster of digital companies in the Temple Quarter Enterprise Zone is using this digital infrastructure to tackle air pollution and traffic congestion, and also to trial self-driving cars.

 KEY TERMS

Place retrofitting Adding important new infrastructure and design features to older places as part of the place-remaking process.

Knowledge economy A mode of production in which greater economic value is attached to the creation of new ideas, innovation, patents and data than it is to physical trade in commodities.

- Glasgow's 'Future Cities' project draws money from the UK Government's innovation fund. The city is pioneering state-of-the-art CCTV, sensor arrays in elderly people's homes and intelligent street lights that dim and brighten in response to movement.
- The small city of Hull, flushed by its success as UK City of Culture 2017 (see page 126), will be the first UK city to entirely abandon its old copper telephone wire network in favour of a superfast fibre broadband network.

There are two points to remember about technology clusters:

- They not only create new employment but also change how a place is perceived by the media and people in other places, thereby driving a virtuous circle of new investment.
- While technology clusters have benefited particular places where they thrive, such as Bristol's Temple Meads, surrounding urban neighbourhoods may gain little benefit from proximity.

CONTEMPORARY CASE STUDY: SILICON ROUNDABOUT

The reimaging of the area around London's Old Street as a smart place is so extensive that some people call it 'Silicon Roundabout'. Its more formal name is East London Tech City (see Figure 4.18). Several major global stakeholders, including Google, Facebook, Intel and McKinsey, have established a presence here, thereby contributing to the place-remaking process. The presence of these transnational corporations demonstrates how Silicon Roundabout has become a powerful place in the context of global networks and systems: anywhere that can attract Facebook and Google must be doing something right in terms of the way it has been rebranded as an investment destination for footloose global capital.

Silicon Roundabout is one of the world's largest media technology clusters measured in terms of new business start-ups per annum. Interestingly, though, this success has been achieved mostly without state assistance and was led instead by players from the private and informal sectors. Ten years earlier, land

prices in Old Street were relatively low precisely because this place had been bypassed by government regeneration schemes that were driving costs upwards elsewhere in London. This attracted young technology and media entrepreneurs looking for cheap rents while trying to build up their businesses.

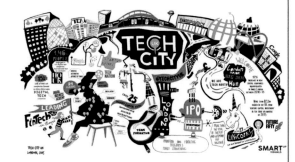

▲ **Figure 4.18** A representation created by a design company to promote places such as East London's Tech City

Smart places in the countryside

Technology has played an important role in remaking rural regions and places too, especially since faster broadband became available in all but the most remote areas. The availability of superfast broadband in southwest England has attracted technology companies and self-employed professionals to Newquay in Cornwall. Magicseaweed, in Devon's Kingsbridge, is the world's largest online surf forecasting platform with 2 million users a month in 2016.

The island communities of Scotland's Inner and Outer Hebrides have been using broadband since the early 1990s to promote economic and social regeneration in a range of ways. The region's primary and tertiary (university) education services have been enhanced greatly by the use of broadband. Migrant 'teleworkers' have chosen to relocate there now that local places are 'switched on'.

In a digitally-connected rural setting it is possible to experience a hybrid sense of place that is both timeless and modern. A traditional landscape of thatched roofs can conceal state-of-the-art technology and business. For many people, this is an attractive combination (see Figure 4.19).

Eco-cities and eco-districts

A sense of being in competition with rival settlements or neighbourhoods often accompanies the place-remaking process. Stakeholders will attempt to position their own home place as the number one target destination for external investors or visitors. One way of achieving this goal is to project an image that is both entrepreneurial and environmentally sustainable.

- A demonstrably low ecological footprint – combined with a commitment to carbon neutrality – can appear impressive to outsiders, thereby making it a valued aspect of place management.
- 'Green credentials' may be an important part of the 'representational regime' that attracts new talent and money to a place (see Figure 4.20).

Environmentally-friendly strategies can be introduced anywhere but it is often harder to retrofit green technology to older structures than it is to apply creative thinking at the planning stage for brand new buildings. Table 4.5 details a range of interesting worldwide initiatives that you could research further.

▲ **Figure 4.19** Throughout the UK countryside, efforts have been made to dig trenches for high-speed broadband cables in order to make 'wild yet wired' places

 KEY TERMS

Ecological footprint The amount of land required to support a place and its people (both in terms of meeting resource needs and disposing of wastes).

Carbon neutrality Achieving net zero carbon emissions using various mitigation measures including energy conservation strategies, 'green' building designs and renewable sources; or by an equivalent amount sequestered or offset.

◄ **Figure 4.20** Manchester's One Angel Square has been the headquarters for the Co-operative Group since 2013 and has a reputation as one of the greenest large buildings in Europe (thanks to a combination of solar power, greywater recycling and natural ventilation)

World region	Measures taken
Tokyo, Japan	The world's largest city by population size, Tokyo has suffered from an extreme urban heat island effect. As a result, many new building developments are covered with shrubs or grass (the vegetation stops heat from penetrating into the urban fabric, thereby lowering urban air temperatures). The Roppongi Hills multiplex cinema complex is covered with a 1300 square metre spread of grass, trees and shrubs.
Mumbai, India	Some new skyscrapers are being designed with a sustainability agenda in mind. One US$200 million, 320-metre tower design stores enough rainwater to service the complex for 12 days, processes its own wastes and features a 215-metre atrium that provides natural ventilation for the building.
Malmö, Sweden	The Augustenborg district is famous for its roof gardens that improve insulation and reduce storm run-off water.
Masdar City, Abu Dhabi	Masdar is a purpose-built eco-city with intelligently-designed buildings, amenities, public transport systems, cycle-ways, water utilities and recycling facilities that contribute to sustainability goals such as carbon-neutral, zero-waste living. For instance, deliberately narrow street design promotes shady conditions and encourages cycling at ground level.

▲ **Table 4.5** 'Eco-city' initiatives from around the world

Carbon-neutral places

Increasing numbers of cities have put carbon-reduction plans into operation since the 2015 Paris Climate Agreement demanded that by 2020 cities reduce their emissions by 20 per cent compared to 1990 levels. In the UK, impressive targets have been adopted by Manchester (Greater Manchester Climate Change Strategy, or GMCCS) and Southampton (Southampton City Council's Carbon Reduction Policy includes a commitment to an 80 per cent reduction in carbon dioxide emissions by 2050).

A place's carbon footprint may also be reduced by producing more food locally, thereby lowering emissions associated with transportation over long distances. Increasingly, small-scale urban agriculture is returning to urban places in the UK on vacant city sites (see Figure 4.21). More people are becoming interested in growing food on allotments too, though far fewer gardens are used for growing vegetables than in the past (for example, during the period of rationing that ended in 1954 and during the 1970s when food price inflation was high).

▲ **Figure 4.21** Urban farming can help (i) reduce a place's carbon footprint and (ii) reimage somewhere by providing it with 'green credentials'

On a more ambitious scale, there is growing interest in using new technologies to enhance urban food production, including hydroponics (growing plants without soil) and aeroponics (growing plants in an air or mist environment). In the UK, this kind of approach to place remaking can be seen in London's Clapham Common neighbourhood, where a cage-like lift takes you 30 m below the ground to 'Growing Underground' – an urban farm housed in a network of dark and dingy tunnels originally built as air-raid shelters during the Second World War. East London's GrowUp Box is an urban aquaponics farm (raising fish and plants together in an integrated system) that was initially bankrolled through the crowd-funding website Kickstarter.

CONTEMPORARY CASE STUDY: REIMAGING THE CITY OF LONDON AS AN ECO-DISTRICT

'The City', or Square Mile, is a small district which is part of the heart of London. Over time, the City district's residential population has shrunk to just 7000 people but around 300,000 people commute into work there every day. This area is the traditional home of the financial services industry, including banks, insurance companies and a range of financial, legal and advisory businesses. These firms have traditionally required a prestigious central London address that is easily accessible for national and international clients travelling to London. A central location also maximises the banks' access to a hand-picked, highly-skilled workforce drawn from the entirety of London's commuting suburbs and the wider southeast region.

The City underwent significant redevelopment following heavy bombing by the German Luftwaffe during the Second World War. Now, this place is changing again: a wave of new architectural additions is helping to bolster the Square Mile's reputation as an eco-district.

Bloomberg's new headquarters is – according to the company's own promotional materials – the most sustainable office building in the world (see Figure 4.22).

■ At a reported cost of £1 billion, water consumption is reduced by 75 per cent and energy consumption by 35 per cent compared with a typical office building. It uses a natural ventilation system – which the architects call 'gills' – to remove heat produced by people and computers in the summer without the need for carbon-emitting air conditioning.

■ The design of the Bloomberg headquarters is intended to be meaningful and authentic too. Its exterior is clad in sandstone and bronze, helping it to blend in with surrounding buildings in central London (where sandstone is a traditional construction material). According to the architects, this was done to make sure that the building was not criticised for failing to appear as a 'London building' despite its technological prowess. In recent years, there has been a backlash against some 'glass box' technoscapes that make absolutely no effort at all to honour local heritage and history.

■ The building was planned and designed by the London architects firm Foster and Partners, who were also responsible for Apple's US$5 billion headquarters in California. This company has become a key global player in place remaking.

Also in the Square Mile, the so-called Walkie-Talkie building incorporates carbon-reducing design measures, including a fuel cell system in its basement that acts like a mini power station. The 37-storey building has a distinctive design – widening as it reaches the top – which led to its nickname. But soon after completion in 2011, the concave shape channelled the sun's rays into a concentrated beam – which newspapers called a 'death ray' – on to the street below, melting car mirrors. For the owners and architect, this was a most unwelcome media representation (the building was designed by another important global player, the architect Rafael Viñoly).

▲ **Figure 4.22** The Bloomberg and Walkie-Talkie buildings are cutting-edge examples of architecture that have enhanced the reputation of London's Square Mile as an eco-district in addition to being a world-class financial hub

Media places

A media place is somewhere we have previously experienced through literature, the screen or art prior to actually encountering it in real life. Our knowledge of a fictional landscape can become blurred with what we understand about the real geography of places used as the setting for a television show or film.

- Locations for films can often prompt a wave of curious ('where was it filmed?') tourism (see Table 4.6). Thanks to their media exposure in the HBO television franchise *Game of Thrones*, several places located north and south of Belfast in County Antrim have been visited by large numbers of the show's fans since shooting first began in 2010–11 (see Figure 4.23).
- In some places, actual building work has been carried out to recreate the film sets visitors hope to see. The Hobbiton Movie Set (from *The Lord of the Rings*) in Matamata, New Zealand first opened in 2002 and at one time received 350,000 visitors a year.
- However, this is not always a very durable form of place remaking: the popularity of some shows and films grows but later vanishes; so too may the visitors.

Film/TV show	Locations
	UK
Harry Potter film series (2001–11)	Alnwick, Northumberland
Peaky Blinders TV series (2013–18)	Small Heath and Deritend, Birmingham
The Wicker Man film (1973)	Creetown, Dumfries and Galloway
	International
The Lord of the Rings and *The Hobbit* film series (2001–14)	Kaitoke Regional Park, New Zealand
The Beach film (2000)	Phi Phi islands, Thailand
Star Wars film (1977)	Chott el Djerid, Tunisia

▲ **Table 4.6** Blurring the line: famous franchises and the real places where they were filmed

▲ **Figure 4.23** The television series *Game of Thrones* has led to the remaking of parts of Northern Ireland as 'media places'. Source: www.fangirlquest.com.

Evaluating the issue

▶ *Assessing the importance of different players in the place-remaking process*

Identifying possible contexts, data sources and criteria for the assessment

The focus of this chapter's debate is the role of different players – alternatively called stakeholders or actors – in the place-remaking process. Which players have the greatest influence? How can their importance be measured?

Before starting to carry out an assessment, it is important to note that place-remaking projects are conducted *at a variety of scales* from street level to the city or even regional level.

- Figure 4.6 on page 124 showed how whole regions and cities of the UK have their own branding slogans; who might have been responsible for creating these important regional messages?
- Plans for the creation of a **Northern Powerhouse** were initiated by the UK Government in 2012; this is an ambitious state-led vision of place remaking on a multi-city scale.
- Alternatively, local neighbourhoods or individual buildings can be renovated or rebranded by local communities acting independently of government or big business.

The assessment also requires that we reflect on what is meant by 'importance'.

- The place-remaking process, as we have seen, is complex and consists of regeneration, redevelopment, reimaging and rebranding. Moreover, each of these actions has a planning stage prior to any work actually being carried out. *Different players may carry out important roles at different times.* One stakeholder – such as a planner or architect – may possess the technical skills and vision necessary to develop a plan of action, but another stakeholder may be responsible for actually footing the bill.

- The examples used below also serve to illustrate the importance of *collaboration* between different players throughout the place-remaking process. Truly ambitious transformations – such as the remaking of large areas of central Manchester and Liverpool – in each case represent around 30 years of hard work by a collective of major stakeholders.

Assessing the importance of governments and organisations

Many of the larger-scale city centre redevelopments which feature in this book would not have happened without the guiding hand of central government. Chapter 3 explained that the challenges for British economy and society posed by deindustrialisation were enormous. Spanning decades, successive initiatives by the Westminster Government have sought to foster the growth of post-industrial economies in the UK's cities and towns, including Liverpool (see pages 124 and 144), Manchester (see pages 121 and 125), Birmingham (see page 160) and of course London (see pages 141 and 158). Some of the most ambitious redevelopment works – such as Canary Wharf in London and the remaking of central Birmingham – have cost many hundreds of millions of pounds.

It is important not to overlook the past role of the European Union in providing much-needed capital for place remaking. For example, during the 2007–2013 European Structural Funds programme, Manchester received £136 million supporting new and existing businesses along with skills and training: schemes that secured EU funding included the National Graphene Institute at the University of Manchester (helping make that part of the city a world-class 'smart place') and the National Football Museum (which received £3.7m).

Local government and city councils play an important strategic role in deciding how such funds as are made available to them are best used. For example, Liverpool City Council (LCC) continues to exert great influence over the evolution of the city's economy and urban fabric.

- Page 124 examined how LCC nominated Liverpool for two global awards as a World Heritage Site (WHS) and European Capital of Culture.
- Most recently, LCC, led by Mayor Joe Anderson, has begun steering the city in yet another direction. In 2011, permission was granted for a futuristic £5.5 billion technoscape development along the city's coastline. The centrepiece of the envisaged new 'Liverpool Waters' development will be the 55-storey Shanghai Tower. The aim is to drive up investment from Asian businesses who, it is hoped, will come to view Liverpool as a desirable locational for their European offices (although the UK's exit from the EU could potentially affect the extent to which these external forces will continue to perceive Liverpool as a true 'world city').
- The city council has also played an active role since 2010 marketing Liverpool to Chinese businesses at the Shanghai Expo annual trade fair in China.

It is important to note, however, that the city council cannot deliver its latest vision for Liverpool without the co-operation of other players. Private-sector companies such as Peel Holdings will pay the massive costs of the actual redevelopment work. The global organisation UNESCO also has a part to play in this drama: it has threatened to remove Liverpool's WHS status if the proposed new waterfront developments spoil the historic skyline of the city. LCC must tread carefully if it does not want the city to be removed from the list of World Heritage Sites.

Another way in which city and county councils can exert influence over place remaking is by campaigning for prestigious awards. Hull gained the title UK City of Culture 2017 after a successful campaign by Hull City Council (see page 126). When it came to delivering the programme of events in 2017, however, Hull City Council set up Hull UK City of Culture 2017 (an independent company and charitable trust) as an independent organisation staffed by people with expertise in the culture sector: the public-sector officials who had spearheaded the campaign to gain the title UK City of Culture were content to take a back seat once it had been won. They engaged others, with greater artistic and arts management expertise, to deliver the actual events. We can therefore offer an assessment that the role of government *may be most important in the early planning stages* of place remaking, whereas the private sector has greater importance during later delivery phases.

Assessing the importance of private-sector players

Private-sector funding, as this chapter has shown, is essential for the delivery of large-scale place remaking. In the example of Birmingham city centre, John Lewis played an essential role in realising the vision of an aspirational new shopping complex.

The private sector as a whole is diverse and has multiple roles to play in place remaking. Private-sector players (at varying scales) include:

- British businesses such as Marks and Spencer, Tesco and Sainsbury's (often acting as anchor stores)
- transnational corporations (whose presence is an important ingredient for the success of new 'retailscapes')
- national and global investors in property and infrastructure (including the sovereign wealth funds mentioned on page 128 and the owners of the UK's various Westfield Centres, including Stratford, London, where an estimated 10,000 people are employed).

UK utility companies also play an important role. Electricity, gas, water and telephone services used to be provided to the general public by national and local government before ownership passed into the private sector during the 1980s and 1990s (today, some UK utility companies are foreign-owned, including EDF Energy, which is a subsidiary of Électricité de France). These businesses have a vital role to play in large-scale redevelopment and regeneration works. For example, an estimated £8 billion was invested by the water company United Utilities in water-quality improvements throughout the River Mersey and the Manchester Ship Canal during the 1990s and early 2000s. In the 1980s, these bodies of water were in a foul state and served as a major deterrent to investment in bankside redevelopment in both central Liverpool and Manchester: United Utilities' investment was essential for regeneration to begin (see also page 145).

Assessing the importance of local communities

Without the involvement of local communities, some attempts to remake places would, of course, undoubtedly fail. Strategies rooted in heritage, art and culture depend on the involvement of local artists. The year of events in Hull when it was UK City of Culture depended on the participation of large numbers of local musicians, writers and artists, including bands The Housemartins and Everything But The Girl and actor Maureen Lipman. A similar story can be told about Liverpool in 2008: The Beatles' Ringo Starr performed at the opening ceremony and Liverpool-born Paul McCartney played a concert in the city later that year, with many more local celebrities in attendance, including footballer Wayne Rooney.

Smaller-scale exercises in place remaking may be entirely driven by local communities. Increasingly, real-world actions are co-ordinated by social media groups composed of local citizens working together.

- Social networks are beginning to play a bigger community-building role, helping to catalyse neighbourhood co-operation and social action.

- In 2011, following disturbances in parts of London, a post-riot clean-up Twitter campaign showed how social media can be used to inspire people to participate in improving their home place's image. In Clapham and Hackney, hundreds of people turned up with brooms to sweep the streets.

At the other end of the scale, however, some extensive redevelopment projects are carried out *in spite of* opposition from local community groups. Examples of opposition to gentrification feature both in earlier chapters and later in Chapter 5 (see pages 168–169). Large-scale redevelopments such as London's Canary Wharf in the 1980s usually necessitate compulsory purchases and demolition of existing housing. In such cases, local communities may become divided. Although some individuals or groups may support change, others rise up against it and attempt to delay work. Dozens of protesters marched against plans to build a McDonald's in East Didsbury in 2017. Members of the 'East Didsbury Not Lovin' It' Facebook group cited childhood obesity, litter, antisocial behaviour and traffic congestion as some of the main reasons to oppose the application (see Figure 4.24).

▲ **Figure 4.24** Local communities play an important role supporting place remaking but they may oppose it too

Assessing the importance of players working in partnership

The examples used thus far help illustrate the very important point that no single player can regenerate or reimage a place by working in isolation. A wide range of expertise is needed, ranging from technical skills to marketing know-how. In the case of large regeneration schemes, the costs may be so high that no single stakeholder can take on the entire financial burden. However, once a wide range of players have strategic involvement in a place-remaking scheme, there is potential for disagreement and delays on account of failure to reach a consensus about what to do and how best to achieve it. You may be familiar with the old saying: 'Too many cooks spoil the broth.'

It therefore follows that any assessment of the importance of different players must also look at the way stakeholders *interact with one another*. A substantial body of academic work (including writing by Michel Callon and Bruno Latour) has explored the way different players become enrolled over time in local 'actor networks'. Callon and Latour's theoretical work offers a framework for understanding how a common vision for place remaking can (or cannot) grow from the interactions of different players. For any scheme to succeed, people with varying opinions need to be prepared to compromise and find common cause. It may be necessary for one player to take a leading role – akin to a team captain or chairperson role – in bringing other players together and helping them to visualise and eventually realise a common goal.

The clean-up of the River Mersey mentioned on page 145 provides a perfect example of this team-building approach. The Mersey and Manchester Ship Canal (a canalised section of the Mersey) flow through the hearts of Liverpool and Manchester respectively. Without them,

neither city would have developed originally. But the water later became the single greatest obstacle to the regeneration and reimaging of both city centres. The manufacturing, chemical and engineering industries that once thrived in Liverpool and Manchester left behind a toxic legacy of heavily polluted water. An estimated billion pounds worth of potential dockside regeneration in both cities went unrealised for many years because of persisting poor water quality.

- One of the worst-affected areas was the foul-smelling Turning Basin at Salford Quays in Greater Manchester (see also page 177). Below the water lay sediments thick with old fat from oil manufacturing and sewage escaped from Manchester's deteriorating Victorian infrastructure. Oxygen levels were depleted by rafts of decomposing sewage-derived sediment and the water sometimes bubbled with escaping methane.
- The challenge was certainly clear for all to see (and smell). But it was less clear where to find the technical and financial support needed to tackle it: the task was so great that it was hard to know where to start.

No single organisation had the money or expertise needed to work alone to restore all of Merseyside's rivers and canals. Ultimately, a large number of private and public organisations shared the effort (see Figure 4.25). This pooling of expertise made problem-solving and decision-making easier. The special role played by the staff of a small government-funded organisation called the Mersey Basin Campaign (MBC) was particularly important. The MBC's role was that of a coach bringing a team of players together. Its action plan put water quality at the heart of the place-remaking process. The two main goals were:

1 improving water quality across the rivers and canals of the Mersey River Basin

2 encouraging and stimulating sustainable urban regeneration by removing riverside dereliction and encouraging attractive new waterside developments as a catalyst for the wider economic regeneration of both main cities.

These goals were ultimately achieved over a 25-year period; the Mersey is cleaner today than at any time since the start of the Industrial Revolution and many of its native fish have returned. Notable major redevelopment schemes that would not have succeeded without the clean-up include Castlefield (see Figure 4.26) and Salford Quays in Manchester, the Albert Docks in Liverpool and the Mersey Waterfront Regional Park.

- The MBC had a crucial role to play in all these successes, which depended ultimately on a cleaner river.
- For one-quarter of a century, the MBC influenced the opinions of the other players, enabled projects to be implemented by groups of stakeholders and mediated between the sometimes conflicting needs and interests of other players in order to build a consensus.

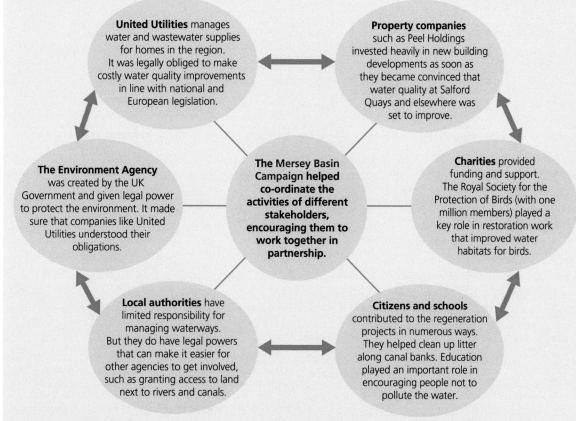

United Utilities manages water and wastewater supplies for homes in the region. It was legally obliged to make costly water quality improvements in line with national and European legislation.

Property companies such as Peel Holdings invested heavily in new building developments as soon as they became convinced that water quality at Salford Quays and elsewhere was set to improve.

The Mersey Basin Campaign helped co-ordinate the activities of different stakeholders, encouraging them to work together in partnership.

The Environment Agency was created by the UK Government and given legal power to protect the environment. It made sure that companies like United Utilities understood their obligations.

Charities provided funding and support. The Royal Society for the Protection of Birds (with one million members) played a key role in restoration work that improved water habitats for birds.

Local authorities have limited responsibility for managing waterways. But they do have legal powers that can make it easier for other agencies to get involved, such as granting access to land next to rivers and canals.

Citizens and schools contributed to the regeneration projects in numerous ways. They helped clean up litter along canal banks. Education played an important role in encouraging people not to pollute the water.

▲ **Figure 4.25** Several important players worked in partnership over 25 years to help improve the River Mersey's water quality and to reimage waterfront places in Liverpool and Manchester. The Mersey Basin Campaign acted as the 'team captain' of this actor network

▲ **Figure 4.26** Waterfront regeneration in Castlefield, Manchester, would have been inconceivable if the canal had not been cleaned first, thanks to the actions of numerous players working together in partnership

Arriving at an evidenced conclusion

Having explored the place-remaking process from the perspective of government, business and communities, it is apparent that their relative importance varies according to (i) the local context, (ii) the scale of the regeneration or reimaging being carried out and (iii) which stage of the remaking process is underway (ranging from planning and funding through to implementation and marketing). In the case of place remaking in Liverpool and Manchester, a strong case can be made that little would have been achieved without the massive financial support provided by United Utilities, a private-sector player. Equally, the guiding hand of local government which propelled both Liverpool and Hull to become cities of culture was of utmost importance in those cases.

Once there is agreement that place-remaking work needs to be done, professional advisers, experts and consultants may temporarily assume a leading role in the decision-making and cost–benefit analysis that is needed to decide the very best course of action. Reflecting on the cultural events staged in Liverpool and Hull, the professional skills of sound, lighting and stage designers were without doubt essential in delivering the final product.

Place remaking is a multi-stage process wherein different players will have a leading role at different times (see Figure 4.27). Ultimately, however, whichever player contributes most to team-building could be viewed as having the greatest importance for the place-remaking process as a whole. It is no easy task to convince many voices to start speaking the same language. The Mersey Basin Campaign, as we have seen, was particularly effective at doing this, helping to build a stakeholder consensus that made the clean-up of the river – and in turn the reimaging and regeneration of two great northern cities – so very successful.

▲ **Figure 4.27** Different players assume a leading role at different times during the place-remaking process

 **KEY TERM**

Northern Powerhouse A plan to increase financial and physical (transport) linkages between the northern cities of Manchester, Liverpool, Leeds, Sheffield, Hull and Newcastle.

Chapter summary

✔ Deindustrialisation and the cycle of deprivation have often created a need for place remaking and regeneration, especially in the inner cities of the UK's largest urban areas, including London, Birmingham, Manchester and Liverpool.

✔ The place-remaking process has many strands, including capital-intensive regeneration and redevelopment, along with the place rebranding, reimaging and marketing work more usually associated with tourist boards but also informal-sector activity (filmmakers, artists and writers).

✔ Place remaking requires collaboration among a wide range of players and stakeholders at varying spatial scales ranging from local community groups to international investment funds and transnational corporations. The interplay between local government, national government and the European Union has been another important aspect of place remaking.

✔ Rebranding and reimaging are increasingly seen as essential tools for successful urban management because of the importance of tourism, leisure and consumer industries in the UK's post-industrial economy: central urban areas suffering from poor image perception are less likely be seen as attractive places for shifting flows of people and investment.

✔ History and heritage industries play an increasingly important role in the British economy and have served as a vital life-support service for many deindustrialised places. Heritage is important for tourism and leisure but is also often the key for manufacturing survival and success: some post-Fordist industries have revived traditional production techniques and heritage products.

✔ Place remaking and regeneration may focus instead on contemporary design and technology rather than dwelling on the past. Some urban and rural places have been redeveloped as technology hubs (smart places) while others have become eco-districts.

✔ Media places blur the line between real life and fictional life: locations that feature as the setting for popular television shows and films gain an opportunity to rebrand themselves for tourism and leisure targeted at fiction fans.

✔ The place-remaking process is complex and may require difficult choices to be made in relation to deciding what is the best way forward for a place and its people. Interactions between different players and stakeholders determine whether outcomes are optimised or not; some players have an especially important role to play in fostering co-operation and shaping a shared vision for place remaking.

Refresher questions

1 What is meant by the following geographical terms? Place remaking; regeneration; reimaging; rebranding; flagship redevelopment.

2 Outline examples of recent flagship development projects in the UK.

3 Using examples, explain the importance of reimaging and rebranding for successful place remaking.

4 Suggest why there is fierce competition between competing settlements for the UK City of Culture award.

5 Using examples, outline ways in which cultural heritage has been used to support new economic activity in different places you have studied.

6 Using examples, compare the characteristics of Fordist and post-Fordist manufacturing.

7 Using examples, outline the characteristics of: urban and rural smart places; eco-cities and eco-districts.

8 Explain why some players have played an especially important role in one or more examples of place remaking you have studied.

Discussion activities

1 In pairs, discuss your experiences of visiting places where large-scale place remaking and regeneration have been carried out. What kind of image do these places project and how successful do you feel any subsequent redevelopment has been?

2 Use Figure 4.3 to isolate different elements of a place-remaking strategy you have studied. Which formal and informal players were responsible for regeneration, reimaging and rebranding?

3 In pairs, create a mind map to show how local, national and global players were involved in one or more examples of place remaking you have studied.

4 In small groups, discuss your personal experiences of heritage tourism in different parts of the UK or other countries you have visited. Have any museums or visitor experiences seemed particularly successful and what would explain this?

5 In pairs, carry out online research about a media place you are both interested in, such as the place where a film or television series you both enjoy watching was made. Look for evidence of visitor attractions and tours. You can also read comments that local people have posted on blogs or websites about the costs and benefits of becoming a media place (perhaps you could even post questions of your own and see if anyone answers them). Once each pair of students has completed their research, they can discuss the findings with the rest of the class.

6 Discuss the importance of different players in the management of your own school; this will help you focus on how different people – including senior management, governors, teachers and students – can all make a contribution to plans for changing places. Imagine that a major decision needs to be taken about, for example: changes in school class size or leaving age; abolishing or changing the school uniform; changing the hours of the school day. Who would need to be consulted? Who would have the expertise to advise on what the possible consequences of that decision might be? Who has the greatest power and would have the final say about what should happen?

FIELDWORK FOCUS

Place regeneration, redevelopment and rebranding strategies lend themselves well to A-level Geography fieldwork and independent investigation titles. There is plenty of scope to produce a study which broadly mirrors one of the more predictable essay titles that deals with this topic, such as: an evaluation of a strategy's success; an assessment of why the strategy was needed; or an analysis of which players were involved.

A *Profiling a regeneration or rebranding strategy and investigating its level of success.* A range of secondary and primary data could be used to document what a particular strategy has achieved (measured in terms of newbuild construction work, landscaping, pedestrian counts, high-street shop profiling and so on). Evaluating whether the strategy has achieved its goals demands a fairly rigorous methodology, however. Secondary research or primary interviews with key players are needed to establish what, if any, goals were set initially. If the purpose of the investigation is to compare 'before' and 'after', this creates the challenge of how to generate primary data showing what the place was like in the past prior to regeneration (and we do not have access to Dr Who's Tardis). One solution is to profile a neighbouring area that is in need of regeneration and contrast this with the regenerated place. This allows samples of primary data to be collected in both locations (environmental quality index surveys can be carried out, along with interviews and questionnaires). The two samples can be compared and contrasted in order to see just how much of a difference regeneration has made.

B *Interviewing key players in order to investigate why a regeneration or rebranding strategy was needed.* This may be difficult to achieve with an extremely large flagship scheme because senior city managers may simply be unavailable for you to talk to. A smaller-scale example of rebranding – such as a rural heritage museum or local urban arts centre – may prove more appropriate as a case study if this allows you to gain access to the actual people who orchestrated the development. You can also profile neighbouring deprived areas that are in need of regeneration and which can serve as a proxy for your own case study prior to its redevelopment.

C *Profiling different organisations, institutions and individuals involved in a place-remaking scheme in order to assess their relative importance in the decision-making process.* You will need to be very well-organised and persistent if you are to succeed in arranging interviews with a significant number of 'persons of influence' – you should probably anticipate sending a large number of emails or making a great many phone calls! Again, it is important to realise upfront that large-scale initiatives (such as Hull's success in applying to become UK City of Culture) may not be very suitable because it could prove extremely difficult to gain access to key players. An in-depth investigation of a smaller, more community-based place-remaking strategy could yield better results.

Further reading

Brown, I. (2017) Coventry named UK city of culture 2021. *The Guardian* [online]. Available at: www.theguardian.com/culture/2017/dec/07/coventry-named-uk-city-of-culture-2021 [Accessed 23 March 2018].

Coe, N. and Jones, A. (2010) *The Economic Geography of the UK* Sage: London.

Devine, T. (2011) *To the Ends of the Earth: Scotland's Global Diaspora 1750–2010.* London: Penguin. GlobalScot.com. Available at: http://www.globalscot.com.

Hall, T. and Barrett, H. (2017) *Urban Geography.* Abingdon: Routledge.

Hartford, T. (2016) Tata Steel, Port Talbot and how to manage industrial decline. *Financial Times* 22 April.

Harrison, R. (2008) *What is Heritage?* (online course) Open University. Available at: www.open.edu/openlearn/history-the-arts/history/heritage/what-heritage/content-section-0?intro=1 [Accessed 21 February 2018].

Perraudin, F. (2017) Liverpool faces up to world heritage removal threat with taskforce. *The Guardian* [online]. Available at: www.theguardian.com/uk-news/2017/oct/03/liverpool-world-heritage-site-threat-taskforce [Accessed 21 February 2018].

Ward, S. (1998) *Selling Places: The Marketing and Promotion of Towns and Cities.* London: Routledge.

Will Self's Great British Bus Journey (2018) [Radio programme]. Radio 4 92–95 FM: BBC.

CHAPTER 5

Creating sustainable places

Place-remaking goals are primarily economic but there are community cohesion, environmental stress and 'liveability' issues to tackle too. This chapter:

- investigates the longevity of government-led economic strategies for places
- examines ways of fostering community cohesion at the local and national level
- analyses symptoms of urban environmental stress and possible solutions
- evaluates strategies to create more sustainable places.

KEY CONCEPTS

Sustainable development Adopted after the 1992 UN Conference on Environment and Development in Rio, the term means: 'Meeting the needs of the present without compromising the ability of future generations to meet their own needs.'

Cumulative causation Self-sustaining 'snowballing' economic growth in a place (involving system growth through the operation of positive feedback). A dynamic and highly integrated set of industries begins to organise around an initial propulsive investment.

Liveability An assessment of what the overall work–life balance in a place feels like, taking into account environmental, community, economic, housing and transport/ commuting conditions.

 # Government policies for economically sustainable places

▶ *How lasting are the benefits of any government efforts to stimulate local economic growth?*

The quest for self-sustaining economic growth

Chapter 4 explored place remaking in a range of local contexts but in each case we cannot know how long-lasting any positive effects will be. Fashions change, once-futuristic technologies become dated and television shows fade from memory over time. Might today's cutting-edge architectural visions one day be reviled the way brutalism is now? Will people still want to see where *Peaky Blinders* or *The Hobbit* were filmed in 2025? Or 2075?

- These are vitally important kinds of questions to ask when assessing the likely success of something. Simply put, place-remaking success is always provisional for each and every case study you come across – we simply don't know how long anything will last.
- Chapter 3 explored a range of emerging economic risks for places, including new technologies and the UK's changing relationship with the EU. External forces and events such as these can change the playing field in sometimes unexpected ways, at which point 'winners' can suddenly become 'losers'.

The economic redevelopment and regeneration of a place therefore requires very careful planning in order to maximise the chance of truly sustainable intervention.

- When it is successful, government intervention results in the growth of a genuine growth pole which (i) serves as a catalyst for wider-scale and lasting economic success and (ii) can display resilience should the wider economic climate worsen.
- In contrast, ill-conceived attempts to revivify economic activity are likely to become derided as a 'sticking plaster' approach to place-remaking policy. In some contexts, flagship developments, sports stadia and other investments have become white elephants (see Figure 5.1).

Olympic legacies

Olympic stadia have a notably mixed track record as economically sustainable venues.

- Many of the venues which cost the Brazilian Government US$12 billion to build and organise for the 2016 summer games, including the aquatics stadium in Rio de Janeiro, are now widely viewed as failed growth poles, or white elephants. The 1997 Summer Olympics facilities in Athens fared similarly poorly.
- In contrast, London's hosting of the 2012 Olympic and Paralympic Games is often seen as the first step taken towards successful regeneration of Stratford in east London. According to this narrative, the UK Government kick-started the transformation of Stratford by funding the construction of sporting venues and, crucially, extensive redevelopment of the local railway station to increase connectivity and capacity. The 2011 opening of the Westfield shopping centre in Stratford with the creation of 10,000 jobs, along with increased demand for housing among professionals who want to move into the area, are evidence of snowballing growth catalysed by public-sector investment in the area's transport infrastructure.

> 🔑 **KEY TERMS**
>
> **Growth pole** A location developed explicitly with the intention of stimulating economic development, employment and the improvement of living standards. Growth poles receive start-up funding as part of a formal government development plan that may involve capital for infrastructure development; private capital is then encouraged to join in. In theory, once established, new investments will attract other industries until self-sustained growth is achieved.
>
> **White elephant** An investment whose cost far outweighs its usefulness.

▲ **Figure 5.1** Before being rebranded as the O2 Arena in 2005, the Millennium Dome in Greenwich briefly suffered a 'white elephant' reputation as an exhibition space that no longer served a real purpose

- In particular, the East Village (formerly the Athletes' Village) has been applauded as a successful attempt at sustainable community building (see Figure 5.2).
- Parts of Stratford remain very deprived, however: wealth and opportunity have yet to trickle down to some neighbourhoods. London Mayor Sadiq Khan observed in 2017: 'There is a big question mark about legacy, or lack of, given the widespread lack of new affordable homes in Stratford'.

▲ **Figure 5.2** East Village in Stratford was purposely designed as a sustainable urban place

The cumulative-causation process

Geographers make use of economic ideas and models to help understand the conditions that must be met for self-sustaining economic growth in local places. The process of cumulative causation is associated with the theories of Gunnar Myrdal, the Swedish economist; Albert Hirschman, an American economist; and John Friedmann, an American regional planner. For many decades, geographers have made use of these authors' original theories and ideas.

- Friedmann's model suggests that the process of cumulative causation explains why growing areas often maintain their initial advantage over other places and continue to develop economically at a fast rate.
- The cumulative causation model sees successfully growing areas attracting even more economic activity directly or indirectly over time. The main reason for this, it is suggested, is that investors, innovators and economic migrants are most likely to be attracted to areas of rapid economic expansion that already possess the appropriate infrastructure, resources and entrepreneurial attitudes (see Figure 5.3).

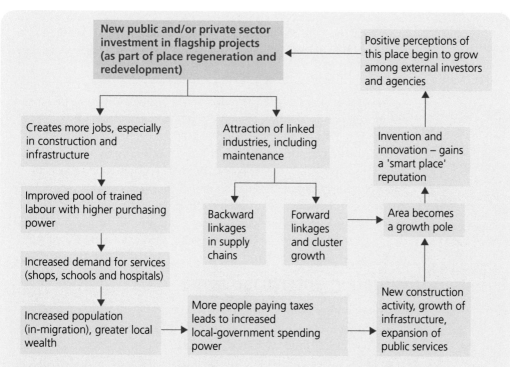

▲ **Figure 5.3** Successful and lasting economic growth relies on cumulative causation (perpetual 'snowballing' effects)

The process of circular and cumulative causation involves several interlinked **multiplier effects**.

1. The arrival of a new and substantial economic activity in an area creates extra employment and raises the aggregate income and purchasing power of a place's population. Immediate virtuous knock-on effects include increased demand for consumer goods and services, schools and healthcare services, and new housing.

2. This, in turn, opens up new avenues of employment for local working-age people, while also attracting professionals to the area.

3. Each new economic activity also attracts linked industries which either form part of its supply chain or use its products. The growth of a cluster of industries in this way leads to further increases in employment and expansion in services and construction.

4. The larger labour pool and expanded local market may require government investment in infrastructure and education services because **demand thresholds** have been crossed.

Thus, new rounds of job and wealth creation continue to appear as part of a virtuous circle of continuing growth and expansion driven by **positive feedback**.

The validity of the cumulative-causation model is, of course, open to debate. Real-world economies are incredibly complex systems and can operate in unpredictable ways.

KEY TERMS

Negative feedback
Changes which help restore lost balance after a system has been altered in some way. If a place 'heats up' economically, negative feedback would involve adjustments that lead to it 'cooling down' again (such as investors losing interest).

Enterprise zone A UK planning tool first introduced by the Thatcher Governments of the 1980s; financial support was offered for businesses in local places designated in need of special assistance.

- An alternative perspective is provided by classical economic theory, which argues that prosperous places will lose their initial advantage over time because of rising labour and land costs. These diseconomies of scale, as they arise, could lead ultimately to the redirection of new investment flows towards alternative, poorer places, thereby evening out the economic landscape once again (an example of negative feedback in a system). According to this view, any growth pole is likely to experience a tail-off in growth over time.
- In theory, a high-achieving growth pole should generate virtuous spillover effects into neighbouring areas also. Myrdal called this a 'spread effect' and Hirschman termed it 'trickle-down'. But it is a contentious economic concept. Many examples come to mind of the way social and spatial growth often becomes polarised over time with little evidence of so-called trickle-down. The persistence of slums in many of the world's wealthiest major cities certainly suggests there are limits to what can be achieved by spread effects from prosperous places.

Evaluating enterprise zones

Some of the examples and case studies in Chapter 4 are too recent at the time of writing for us to offer a meaningful long-term assessment of their success. Insufficient time has passed since Hull's year as UK City of Culture 2017 to know the extent to which it has improved the city's long-term economic and social fortunes (even if the early signs are promising). In contrast, it is possible to offer a proper retrospective evaluation of the UK enterprise zones that were established for the first time in 1981. Designated as being in need of special assistance, these were usually inner-city areas where the worst manufacturing and mining job losses had been experienced.

- New firms investing in these places were exempted from paying local taxes for a decade.
- Each enterprise zone was administered by an Urban Development Corporation (UDC), such as the London Docklands Development Corporation (LDDC).
- UDCs were given the power and money to acquire and reclaim brownfield sites. They also played an important role in developing local infrastructure to help 'pump-prime' these sites for rounds of new investment from the private sector.
- In principle, enterprise zones functioned as growth poles, the intention being that wider spatial regeneration would take place beyond the margins of each designated area. Trickle-down or spread processes would, it was hoped, infiltrate their way into surrounding districts of cities.

Twenty-five enterprise zones were designated in two rounds between 1981 and 1984 for economically hard-hit parts of Clydebank, Corby, Dudley, Hartlepool, Salford, Speke, Swansea, Tyneside, Wakefield, Allerdale, Delyn, Glanford, Invergordon, Middlesbrough, Milford Haven, northeast Lancashire, northwest Kent, Rotherham, Scunthorpe, Tayside, Telford, Wellingborough, Belfast, Londonderry and Isle of Dogs (London

Docklands) (see Figure 5.4). Did they all work? Critics of this first wave of enterprise zones have argued that many did not prove durable while others leached employment and investment from neighbouring areas in what became a 'zero-sum' game. This meant the enterprise zone only made gains when neighbouring places suffered losses – the opposite of a hoped for trickle-down effect.

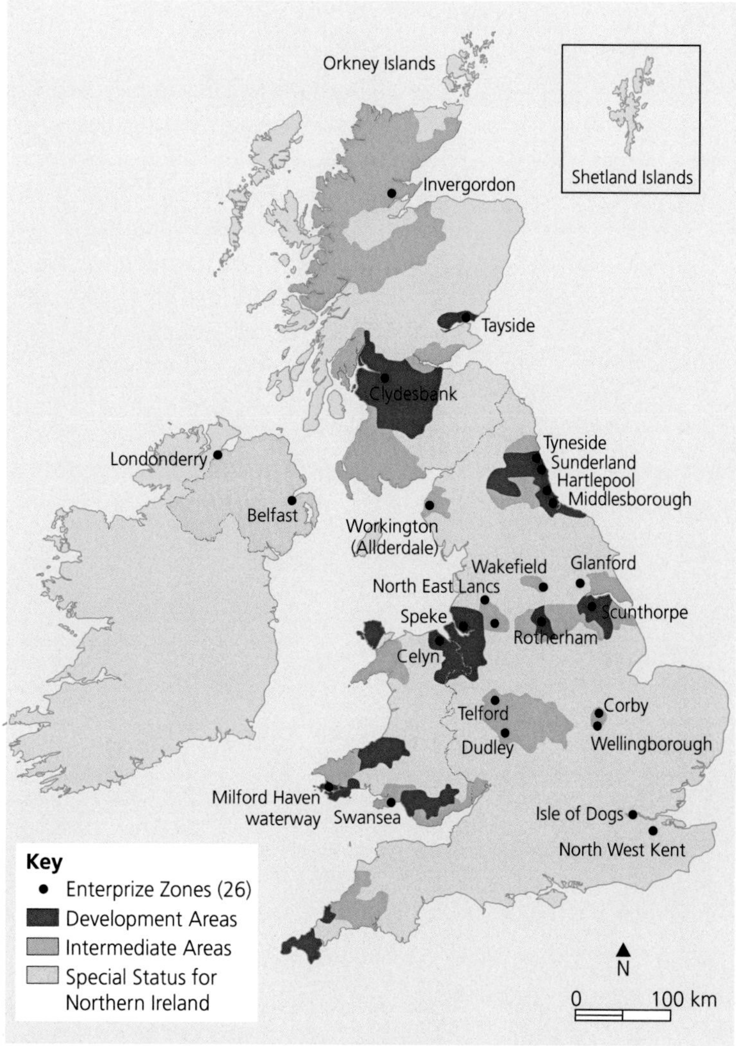

- According to Centre for Cities, enterprise zones in total gained 58,000 jobs, *but over 40 per cent of these were associated with businesses that had relocated to enjoy the tax cuts.* The 'zero-sum' critique therefore appears well deserved: one place's success came at a cost to another. This meant the total net employment benefits of the enterprise zones were relatively small.
- Enterprise zones were expensive to deliver as well. On average, one additional job in each enterprise zone cost the Government £26,000 in today's prices.
- Moreover, the benefits were not always long-lasting. Some of the places mentioned above – including Wakefield and Middlesbrough – performed relatively poorly in the *Cities Outlook 2018* report by Centre for Cities.

▲ **Figure 5.4** Enterprise zones were designated in many places, including Wakefield, Dudley and Scunthorpe, in the 1980s. But how permanently did they change the UK's economic map?

This critique does not fully apply to the well-known Isle of Dogs (London Docklands) enterprise zone though. Over the last 35 years, this place has been redeveloped into a financial hub of unparalleled significance within Europe. It has also served as a growth pole around which layer after layer of redevelopment and regeneration has built up in the surroundings of east London. This success has, however, been accompanied by rising levels of inequality that previous generations had sought to eliminate. An enormous wealth divide now exists between the highest-class housing areas and some poorer areas around the edges of Isle of Dogs (suggesting that the trickle-down process has not always benefited neighbouring places and communities greatly).

An overall evaluation of the 1980s enterprise zones strategy might therefore conclude that:

- in most places, only modest economic gains were made, sometimes at the expense of neighbouring places
- the benefits were often short-term only – when the ten-year funding period came to an end, some businesses closed or moved on, notably so in Corby and Hartlepool
- the Isle of Dogs (London Docklands) was a great success when measured in economic terms, but there were social costs for local people displaced by the redevelopment.

CONTEMPORARY CASE STUDY: THE ISLE OF DOGS (LONDON DOCKLANDS)

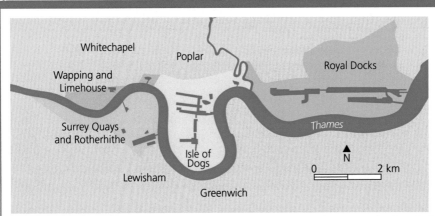

▲ **Figure 5.5** The Isle of Dogs is land enclosed on three sides by a meander bend of the River Thames

July 2017 marked the thirty-fifth anniversary of the establishment of the Isle of Dogs enterprise zone, sited in east London's Docklands area (see Figure 5.5). A place that had suffered from dereliction, decay and rising unemployment was, by the 1990s, utterly transformed thanks to an ambitious development plan that strategically pursued long-term regeneration goals by attracting a new generation of 'post-industrial' employers (service-sector businesses).

In 1981, the London Docklands Development Corporation (LDDC) was set up with a daunting mission of regenerating hundreds of acres of industrial wasteland. Large manufacturing employers, such as textile and food-processing industries, began closing their doors in London after the 1960s on account of global shift and changes in container shipping technology. Newly-designed transport ships grew in size to carry larger containers in ever-greater numbers. However, larger vessels were unable to travel inland along the Thames as far as the Isle of Dogs. Instead,

new docks were built downstream in deeper water, closer to the North Sea, at Tilbury. More inner-city dock closures followed for London.

The Isle of Dogs also suffered from:

- deteriorating housing stock built in the 1800s (one-third of housing in London's Docklands was considered unsatisfactory for habitation in 1981)

- building dereliction and ground pollution left in the wake of deindustrialisation (by 1981, 50 per cent of Docklands was classified as derelict)

- out-migration of people who could afford to (they generally shifted towards London's suburbs, leaving behind a vicious circle of poverty; in 1981, unemployment in the Isle of Dogs area had reached 24 per cent, twice the national average).

For almost 20 years, the LDDC worked to turn around the fortunes of this place in partnership with other players, including the UK Government and property developers who would ultimately spearhead the large building projects in the Isle of Dogs. The LDDC paved the way for the Docklands Light Railway (DLR), London City Airport and the Jubilee Line rail extension (and Canary Wharf station). Few people in 1981 would have believed that the Isle of Dogs would become such an economic powerhouse and cultural hub, with a rapidly-growing residential population.

Since the mid-1990s (prior to which growth was slow) the Isle of Dogs has gone from strength to strength, overtaking the City of London district (see page 141) in 2012 to become the biggest urban employment zone in London for banking and financial services. The six biggest banks in the UK now jointly employ almost 50,000 workers in a cluster around Canary Wharf station. HSBC and Barclays are headquartered there; so too is Credit Suisse's investment banking operation and the US banks Citigroup and Morgan Stanley.

This place's success is explained in part by the large-capacity high-rise blocks which were built (see Figure 5.6). This has allowed large financial businesses to house all their staff on a single site (elsewhere in London, large businesses typically need to acquire an entire suite of Victorian or Georgian buildings, each with far fewer floors than the Canary Wharf skyscrapers). Another reason lies in the nature of its connections with other places and financial markets. Canary Wharf offices are in the perfect time zone for business and trading by bridging the time gap between the USA, Japan and, increasingly, China. The English language is another important local asset explaining London's historic role as a global financial centre.

Overall, the Isle of Dogs has performed through time in the way cumulative-causation theories predict successful growth poles should do. Studies show £10 of private investment made for every pound of public money spent, providing clear evidence of powerful multiplier effects.

Other assessments have been more critical though.

- When work first began on remaking the Isle of Dogs into a global financial hub, it became clear that many existing community organisations were strongly opposed. In 1985, community groups organised a flotilla called 'Docklands Fights Back'. Despite a few victories, however, such as the Cherry Gardens estate (which was retained by Southwark council for local people), the revivified Docklands ultimately offered little to the area's unemployed communities.

- Very few of the new jobs created since 1981 went to those families who lost their jobs when the docks of London closed. Local people who did get work often found themselves occupying menial and low-paid roles, such as cleaners or baristas.

- The new housing was expensive and soon forced up rents and mortgages in the surrounding areas.

It remains to be seen how resilient this place will be in the event of a 'hard' divorce between the UK and the EU. Some international banks may choose to relocate to continental Europe after Brexit (see page 92) unless a deal can be struck which allows flows of workers and money to move relatively freely between Canary Wharf banks and other EU financial hubs such as Frankfurt. At the time of writing, it is far from clear whether this can be achieved and what will happen if it cannot.

Twenty-first century enterprise zones

A new wave of enterprise zones were designated by David Cameron's 2010 UK Government. Between 2011 and 2017, 48 local areas were chosen. In contrast with the 1980s – when severely-struggling problem places were designated – areas with strong medium-term growth prospects were prioritised in order to give this next generation of growth poles the greatest chance of success.

- In particular, the 2010 recipient regions had already experienced some modest success prior to selection, often in the quaternary (technology) sector.

- Examples include the North Kent Enterprise Zone (Ebbsfleet), Dudley's Business and Innovation Enterprise Zone (a new technology hub for the Midlands) and Cornwall Marine Hub (focused on marine renewable energy at Falmouth Docks).

▲ **Figure 5.6** The financescape of the Isle of Dogs

Only a handful of new enterprise zones were picked by central Government as top-down choices. The majority were chosen instead with bottom-up input from community stakeholders belonging to local enterprise partnerships (LEPs). The LEPs are bodies composed of representatives from local authorities, universities, businesses and voluntary-sector players. In some cases, neighbouring LEPs have worked together to identify a single merit-worthy growth pole in their wider region, thereby fostering co-operation between different places and communities. The selection criterion used was always 'the place most likely to succeed' rather than 'the place most in need of help'.

Businesses basing themselves in new enterprise zones have been able to access a number of benefits:

- a business rate discount worth up to £275,000 per company over a five-year period
- generous enhanced capital allowances (tax relief) worth millions to businesses making large investments in buildings and machinery.

Between 2012 and 2015, the UK's new enterprise zones reportedly attracted over £2 billion in private investment.

CONTEMPORARY CASE STUDY: BIRMINGHAM CITY CENTRE ENTERPRISE ZONE

Central Birmingham gained enterprise zone designation in 2011 (see Figure 5.7). This place is now managed by an alliance between Birmingham City Council (BCC) and the Greater Birmingham and Solihull Local Enterprise Partnership (GBS LEP). To date, they have overseen:

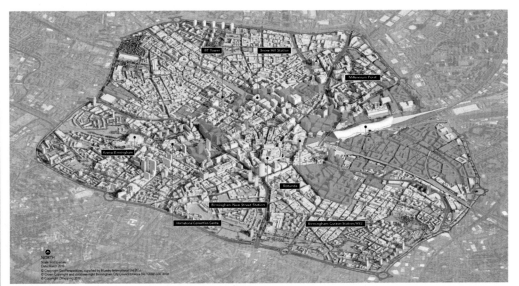

▲ **Figure 5.7** The new Birmingham City Centre Enterprise Zone is made up of 39 sites spread through the heart of the city

- £125 million spent on upgrading city infrastructure, broadband and office space
- the £600 million construction of the Grand Central Shopping Centre retail zone above the refurbished Birmingham New Street Station, with John Lewis as the 'anchor' tenant (shown in Figure 1.11, page 20).
- the linking together of the city centre's different railway stations at a cost of over £100 million

In a joint statement, the LEP and BCC have set themselves a benchmark for further success: 'By 2031 Birmingham will be renowned as an enterprising, innovative and green city that has undergone transformational change growing its economy and strengthening its position on the international stage.' In 2017, it was also announced that Birmingham will host the Commonwealth Games in 2022, which will be the largest-scale sporting event to be staged in the UK since the 2012 Olympic and Paralympic Games.

Although it is too early to judge whether Birmingham City Centre will enjoy the same lasting success as the Isle of Dogs (see page 158), its future prospects appear good. This is, perhaps, to be expected given that this place was carefully selected by local players on account of its high prospects for growth if provided with extra support.

2 Encounters with cultural and social differences and inequalities

▶ *How can cultural and social issues be managed to provide local communities with a more cohesive and sustainable future?*

The lived experience of cultural diversity

Chapter 3 explored how cultural diversity has grown over time in the UK, in turn requiring policy-makers to put greater thought and effort into the careful management of minority ethnic group welfare and community cohesion for society as a whole. Alongside sustainable economic growth strategies, the creation of liveable places often depends also on successful management of social and cultural tensions, divisions and inequalities. Where there is good governance, multiculturalism can result in diverse and vibrant heterogeneous neighbourhoods (see page 105). Prejudice and a lack of mutual understanding has sometimes brought tension and conflict to certain places, however, particularly whenever cultural diversity issues have become interlinked with a serious economic challenge, such as deindustrialisation.

At a local level, UK schools and colleges are important places where personal identity is negotiated and explored by young British people of increasingly varied heritage and ethnicity. For example, at Bancroft's School in Essex, a large mixed body of Muslim, Sikh, Hindu and Buddhist students collaborate with their White British peers each year in the Taal festival.

Traditional and modern music and dance combine to provide a complex cultural expression of what it is like to grow up with one foot in London life and another in Asian family culture. In 2016, the students performed a Bollywood-style update of Jane Austen's *Pride and Prejudice*.

CONTEMPORARY CASE STUDY: POSITIVE ENCOUNTERS WITH DIVERSITY IN NEWHAM

Newham is an east London borough where levels of diversity are higher than anywhere else in the UK. Thirty-six per cent of all births in Newham during 2016 were to foreign-born mothers. Indians, Pakistanis, Filipinos, Somalis and more recently Poles and Romanians live in large numbers in Canning Town, Plaistow and Barking Road.

One 2016 newspaper report used a positive tone when portraying highly-diverse Newham as 'an arrival hall to the world'. There is an endless procession of different ethnic food shops. Travel agents advertise cheap flights to Kingston and Dhaka, while leaflets in shop windows advertising rooms to rent are written in 40 different languages. Twice every day, a local citizenship ceremony for Newham residents is presided over by local councillors in a room crammed with representations of British life including Union Jacks and a portrait of Queen Mary.

For more than a decade, Newham council has striven to improve integration. Free English classes are offered to recent arrivals. There is also a broader strategy to invest in public gardens, libraries, street parties and volunteering networks in the hope that people from diverse communities will come together. It is not easy to collect qualitative data supporting whether or not such a policy works because views can be subjective on what constitutes 'success'. But a recent liveability survey showed that nine out of ten residents who were interviewed believe Newham is a place where people from different backgrounds 'get on well together' – something which many young people now take for granted.

Critics say that one community that has been left behind is the White British who once were the borough's majority population. Between 2001 and 2011, the proportion of white people fell by a half and in some wards they now account for less than five per cent of the population. Those that remain tend to be in low-income groups.

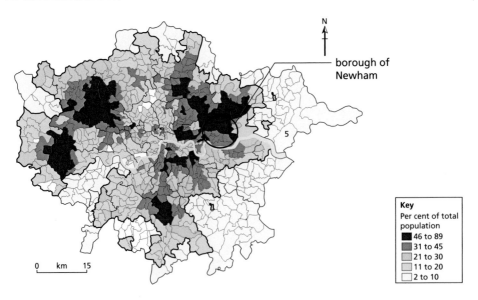

▲ **Figure 5.8** The distribution of minority ethnic groups in London in 2010, with the borough of Newham highlighted

Causes of cultural and social tension

During the 1970s, racial tensions in British inner cities were framed by the general crisis of capitalism which the UK and other major industrialised powers were experiencing. Rising unemployment after 1976 was experienced most acutely in the country's industrial heartlands where large numbers of Black and Asian post-colonial migrants had arrived in the previous decade (see page 107). Right-wing political movements such as the National Front played a disingenuous 'blame game' by asserting that unemployment was a result of immigration (its true causes were global shift and a cyclical slowdown in the world economy triggered by rising oil prices). In 1982, the writer Paul Gilroy argued that a 'new racism' had become interlinked with 'the new realities of structural change and economic uncertainty'.

During these years, tensions in some places amplified into violent disorder and conflict with police forces (see Figure 5.9). There was rioting on the streets of Toxteth (Liverpool), Brixton and Broadwater Farm (London) and Chapeltown (Leeds) during the early 1980s. Other affected cities included Bristol, Sheffield, Manchester and Birmingham. The causes of the rioting were complex but included 'stop and search' tactics used against young Black people. But the Brixton and Toxteth riots also coincided with the highest unemployment figures the UK had ever experienced. Deindustrialisation disproportionately affected the UK's relatively new Black and Asian communities, whose as-yet limited social mobility left them highly vulnerable to manufacturing job losses.

One conclusion about the causes of community tension at this time is therefore that underlying structural economic problems were the real cause, rather than cultural issues. Continuing racial tensions in many North American inner-urban areas are undoubtedly also linked with the economic challenges left by the deindustrialisation of once-great cities such as Detroit, Baltimore and Jackson. Powerful cultural divisions and inequalities in US society persist in places where structural economic changes have pushed many African-American families over the poverty line.

Rioting happened again in British cities in 2011. The timing was significant. It was three years after the height of the global financial crisis and one year since the UK Government had introduced its so-called austerity measures (see page 87). But house prices in southern England and some major cities were already growing once again, increasing the wealth of those lucky enough to own them. On this occasion, a broad mix of young people (both white and minority ethnic) were involved, suggesting that the cause of rioting had less to do with the grievances of any one particular ethnic group than it did with social and economic pressures experienced by poorer young people in general.

▲ **Figure 5.9** Rioting in Brixton in 1981

▲ **Figure 5.10** The BBC cast a British actor with Guyanese heritage, Angel Coulby, as Queen Guinevere in its television series *Merlin*. The myth of King Arthur has long been regarded as an important chapter in 'the story of Britain'

Progress in the management and representation of diversity

The widespread rioting which took place in UK cities around 1980 prompted an influential inquiry called the Scarman Report which subsequently had an important role in highlighting the long-term need to remake and rebrand inner city places in order to (i) tackle social problems linked with the inner-city cycle of deprivation and (ii) improve race and community relations. Since then, a raft of measures have been introduced to tackle inequality and discrimination, and to foster community cohesion.

● Table 5.1 shows a range of initiatives and increased efforts to represent places and communities in the UK as diverse, multicultural and inclusive.

● The national curriculum for schools is also important for the delivery of this outcome. Teaching and learning about the experiences, beliefs and perspectives of different groups in society, delivered through subjects which include Citizenship and Geography, plays an important role in building mutual respect and understanding.

Area	Examples of multicultural integration
Media	In 2016, the BBC introduced a diversity and inclusion target of fifteen per cent for Black, Asian and minority ethnic (BAME) people appearing in its programmes on screen and in lead roles. Even in historical representations of places, viewers can expect to see a diverse community represented (see Figure 5.11). In the television series *Doctor Who,* the fictional character Bill Potts was a gay Black woman. Her portrayal as a charming girl-next-door character helps community cohesion by illustrating the normality of diverse sexuality.
History	Black history – and Black people's contribution to the narrative of national history – is now taught in British schools. The BBC in 2017 produced a series *Black and British: A Forgotten History* in which David Olusoga unveiled commemorative plaques in places of significance in the UK, Africa and the Caribbean. One plaque in Rochdale tells the tale of the nineteenth-century mill workers who showed solidarity with those in slavery by refusing to handle cotton picked by slaves in the southern states of the USA.
Sport	Twenty-five per cent of British football players now come from a non-white ethnic background. Cyrille Regis was one of the UK's first Black players, who played for England five times during the 1980s. He spent seven years at West Bromwich Albion, during which time, despite suffering horrible racism from the terraces, he played on undeterred. Regis was a role model and source of inspiration to thousands of children.
Culture	In 2016, British actress Sophie Okonedo played Queen Margaret in the BBC's production of *The Hollow Crown*, an adaptation of Shakespeare's *The Wars of The Roses* plays. David Oyelowo was the first Black actor to play a white monarch in a major stage production in 2001 when he took on the role of Henry VI for the Royal Shakespeare Company. By using Black actors in these roles, production companies are reflecting the community make-up of places in the UK today.
Politics	In the 2015 election, 3 million votes were cast by members of minority ethnic communities, up from 2.5 million in 2010. The 2017 General Election brought in more BAME MPs than ever before, with the number increased from 41 to 51, including the first female Sikh MP (Labour's Preet Gill). In 2018, high-profile politicians of non-white backgrounds included Home Secretary Sajid Javid, Shadow Home Secretary Diane Abbott, David Lammy and the London Mayor Sadiq Khan.

▲ **Table 5.1** The UK is increasingly represented as a diverse society

Differing perspectives on managing diversity

On buses and trains in most major cities of the UK, a variety of spoken languages can routinely be heard. Many places have been transformed culturally as a result of the postwar migration processes outlined in Chapter 4 (see page 106–107).

Inevitably, views on these changes differ.

- Some people judge the scale and rate of cultural change to have been too great. Many of those who voted in favour of the UK leaving the EU in 2016 did so because of their disapproval of high levels of immigration into the country in recent years. In their view, the issues of immigration and cultural change have been poorly managed by successive UK Governments.
- Other people view the UK as a successfully-managed multicultural society where migrants are welcome and Muslim women should be able to wear the traditional face-covering niqab veil in public if they choose to.
- A third view is that international migration is economically desirable but that increased ethnic diversity should be managed in ways which minimise cultural changes to places. For this to be achieved, so the argument goes, minority ethnic communities should do more to assimilate into a British way of life. However, past arguments made in favour of assimilation – including the suggestion that British Asians should be supporters of the English cricket team – could offend a present-day audience (see Figure 5.12).

In 2016, the Casey Review accused the UK Government of 'a failure of collective and persistent will' in its efforts to improve community cohesion, however. The report pointed to the risks associated with growth of isolated and ghettoised communities in recent years.

- Acts of violence across Europe have made it hard to judge the management of urban cultural diversity as having been entirely successful. Terrorist attacks carried out by British citizens caused the deaths of 56 people in London in 2005 and a further 22 in Manchester in 2017. As a result, steps are now being taken to foster community cohesion in local places where the risks of radicalisation are deemed to be high (Figure 5.11)
- Many radicalised (and some would say vulnerable and impressionable) young Muslims have travelled from the UK and other European countries to fight for Daesh (or

▲ **Figure 5.11** The Westfield shopping centre in Stratford, London: steps are being taken to strengthen local community cohesion within this globalised retail space

 KEY TERM

Assimilation The process by which incomer groups adopt some of the values and norms of wider society.

Today

The ultimate test for being British: Which side do the Asians cheer for at cricket?

TEBBIT RACE BOUNCER

'Which side do they cheer for?' the former Tory Party chairman asked.

NORMAN TEBBIT bowled the Government a vicious bouncer yesterday when he claimed many Asian immigrants failed his 'cricket test' to be a true Brit.

▲ **Figure 5.12** This 1990 newspaper provides qualitative source material for critical analysis. What does it tell us about debates surrounding national identity and community cohesion back in the 1980s and 1990s? How has the debate moved on since then?

so-called ISIS) in Syria and the Middle East. The return of Jihadists to the places they left previously is of particular concern to security services in the UK and other European countries.

- Acts of violence have been carried out by individuals in other contexts too: MP Jo Cox was murdered in 2016 by a radicalised member of a group with far-right ideological beliefs.

Some critics of the Casey Review were left uneasy by its 'assimilationist' tone. However, cultural diversity clearly continues to create new challenges alongside the positive changes that many people feel it brings to places.

Managing gentrification and social inequality

When affluent people and new investment flow into a place, gentrification occurs. This at-times complex process has helped reshape places throughout the UK in recent decades, including now-fashionable gentrified districts like London's Notting Hill and Hoxton. It involves new flows of people and investment into previously neglected or declining places (see Figure 5.13). The word entered the vocabulary of geographers over half a century ago, having first been introduced by the British sociologist Ruth Glass in 1964. Important early geographical research into gentrification was carried out by Neil Smith. Along with many other urban and social geographers, Smith critically examined the rapid gentrification of cities throughout the developed world during the 1980s, partly as a result of housing market deregulation. Another important factor driving gentrification was the growing importance of property investment as a post-industrial economic activity.

Gentrification is often integral to place remaking. The process tends to be complex and lends itself well to critical thinking about the important geographic concept of positive feedback. This is because gentrification has both people-led (or bottom-up) and institutional (or top-down) causes.

🔑 **KEY TERM**

Creative class A demographic group made up of knowledge (quaternary-sector) workers, artists (including musicians, actors and so-called 'hipsters') and intellectuals (including university staff and journalists). The academic Richard Florida has written at length about the contribution of the creative class to successful and sustainable place remaking.

- It is people-led in so far as investment and renewed interest in old neighbourhoods usually begins with the arrival of a **creative class** of artists and young (thus still relatively low-paid) professionals who have been searching independently for an affordable and central place to live that has 'character' and 'authenticity' (unlike their perception of the suburbs). Rundown areas which have undergone deindustrialisation may

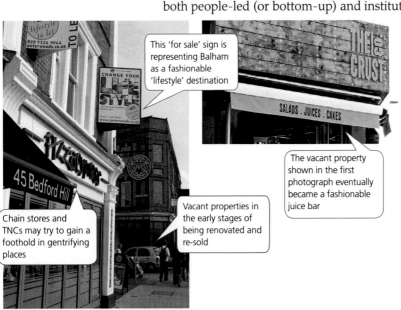

This 'for sale' sign is representing Balham as a fashionable 'lifestyle' destination

The vacant property shown in the first photograph eventually became a fashionable juice bar

Vacant properties in the early stages of being renovated and re-sold

Chain stores and TNCs may try to gain a foothold in gentrifying places

▲ **Figure 5.13** Field evidence of gentrification in Balham, London nonetheless retain interesting and attractive heritage and architectural features which the creative class are attracted to.

- Gentrification is driven by top-down investment too. The construction of a new train, tram or tube station can revitalise an old neighbourhood, sending property prices spiralling upwards. Working together, bottom-up and top-down forces co-create a linked cycle of positive feedback. Each new flow of migration and investment leads, in turn, to further inflows of money and people (see Figure 5.14).

- It is not an exclusively urban process; rural places can become gentrified too. House prices in some Oxfordshire villages and towns have ballooned in recent years and a four-bedroom house in the small country town of Thame cost around £500,000 in 2017 (around twice the national average).

The benefits and costs of gentrification

'Community workers, policymakers and urban planners all across the world are grappling with the challenge of integrating old and new communities and building a sense of place.' [*Financial Times*]

Gentrification is a place-remaking process with varying benefits and costs for different individuals and societies. It is usually linked with the (re)creation of wealthier neighbourhoods. Property prices rise when demand for housing increases in a neighbourhood. Previously low-worth housing stock – which had been rented cheaply to lower-income groups – is sold to private buyers once its value starts to rise. Land values in inner London's most gentrified postcodes reached an all-time high between 2014 and 2016.

By this measure, place remaking is deemed 'successful' when an area becomes a property hotspot. Significant demand for housing drives land values upwards, at which point large-scale inward investment by property developers and retailers can be expected. Inevitably, there are adverse effects for poorer longer-established community members though.

- Younger members of some local multigenerational families may have fewer skills and earn less compared with professional incomers; they cannot afford to leave the family home and buy their own property locally.

- Low-income tenants are evicted and must find new homes elsewhere instead, especially where economic overheating occurs, resulting in displacement, so-called 'decanting' and community disintegration (see Table 5.2).

- These processes can be seen operating all around the UK, from St Ives and Plymouth to Salford and Peckham.

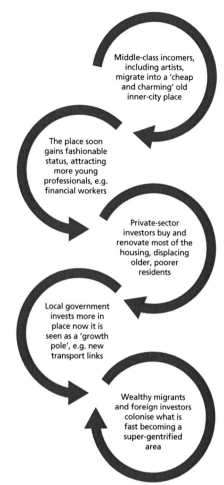

▲ **Figure 5.14** How positive feedback creates a 'virtuous cycle' of migration and investment flows into an inner urban place (but also displaces older residents)

▲ **Figure 5.15** Chelsea Harbour in London: this is a gated neighbourhood characterised by high security, foreign ownership and limited interactivity with other surrounding places

KEY TERM

Gated neighbourhood
A high-class, securitised residential development with fixed entry points. Some gated places make use of surveillance cameras and security guards.

The negative impacts of gentrification are very much another example of a 'wicked problem' (see page 86) in the study of Geography.

New **gated neighbourhoods** are sometimes created by developers in gentrifying areas. These have been strongly criticised for their reluctance to integrate new residents and the surrounding community. Chelsea Harbour in London (see Figure 5.15) has been criticised as a 'mono-use' development which makes no effort to connect with the neighbourhoods that surround it and denies outside people the opportunity to interact with it as a place. It is also unclear how internally cohesive a place like Chelsea Harbour will ever become. One hundred years ago, a church or community centre would usually sit at the heart of a neighbourhood. Today, new residential developments like Chelsea Harbour can offer only a gym or small supermarket as a neighbourhood core. Is this sufficient to help shape a sustainable sense of place and identity?

Issue	Impacts
Overheating	Part of the difficulty in managing places sustainably is the way that economic overheating can occur in successful areas. For example, in Hackney, London, house prices rose by a scorching 65 per cent between 2012 and 2017. ■ A common dilemma quickly arises for people working in creative industries, such as artists and writers. No sooner have they helped provide a newly gentrified place with its creative 'vibe' than landlords begin to raise rents, thereby forcing the artists to move on. In London, the cycle was seen in Notting Hill in the 1960s, Shoreditch in the 1990s and Hackney in the 2000s. ■ Some **super-gentrified** urban areas have become unaffordable for **key workers** such as junior teachers and nurses. The implications for community sustainability are alarming. ■ As a result, planning officers will often insist that new property developments include some affordable housing. However, critics say affordable housing makes up only a very small percentage of housing stock and cannot compensate for more widespread loss of cheap accommodation. ■ Overheating can harm small businesses too. Recently, business rates across London were increased for the first time in seven years, with the borough of Hackney experiencing the biggest increase of almost 50 per cent. For small hipster businesses who helped drive the regeneration of Hackney, the rise has made it impossible for some to make a profit.
Community disintegration and decanting	Gentrified areas are often represented as cohesive places where local people get together in new bars, restaurants, markets and at community events. ■ But low-income communities may become displaced – or decanted – from gentrified places over time. This involves their relocation to another area entirely or, in the case of London's Grenfell Tower, into poorly designed and hazardous social housing (see page 111). ■ The £8 million Earls Court redevelopment scheme in London is at the heart of a dispute that began in 2009. Plans for 7500 mostly expensive homes – involving the demolition of two council estates and the rehousing of 760 existing low-income households – were initially backed by local councils and the London Mayor. But campaigners against the redevelopment say that insufficient affordable housing will be provided as part of the works. The result is a fiercely-divided community. ■ Gentrification has a cultural dimension sometimes; lower-income ethnic groups have been largely displaced from some parts of Balham (see Figure 5.14), Tooting and Brixton in London. African-Caribbean ethnic food and hair product shops have given way to new bars and restaurants aimed at a different clientele. ■ This is what makes gentrification a contested and political process; the resulting tensions may build into more serious conflict and crime, as happened with the Cereal Killer Café in London's Brick Lane.

Table 5.2 The linked issues of overheating, community disintegration and decanting

KEY TERMS

Super-gentrified (or hyper-gentrified) An excessively-gentrified place, often as a result of private and public investment reinforcing one another through positive feedback.

Key workers People whose work provides a vital community service (such as police, health workers and teachers).

CONTEMPORARY CASE STUDY: THE CEREAL KILLER CAFÉ

Alterations to the built environment brought by gentrification may benefit some groups but can provoke hostility from others who perceive place changes (and those responsible) negatively.

Alan and Gary Keery founded the Cereal Killer Café in London's Brick Lane. It sells bowls of cereal for £3.50. In 2015, the Keery brothers' business was targeted by a crowd of anti-gentrification rioters. According to *The Guardian* newspaper, the protest, which had been organised online by the anarchist group Class War, became focused on the café as 'a symbol of inequality in east London'. The crowd – some of whom carried pigs' heads and torches – sprayed the word 'scum' in red paint on the café's front door.

- In an interview with London's *Evening Standard* newspaper, Alan Keery said: 'The protesters knew exactly what they were doing. They knew if they went for us they would get media attention. It's clever of them.' He and his brother claimed they had also received death threats.

- The protesters told reporters they were angry about neighbourhood changes. The issue of affordable housing was foremost in their minds, along with the higher prices now being charged by retailers, bars and cafés.

- Previously, a *Channel 4 News* presenter had criticised the Keerys for selling over-priced cereal in a place with high rates of poverty.

- Since the events of 2015, the brothers have expanded and globalised their business, setting up Cereal Killer branches in Birmingham, Kuwait and Dubai.

▲ **Figure 5.16** The Cereal Killer Café

Keeping the creative class

One London borough, Barking and Dagenham, recently introduced policies designed to help keep its creative class. The council gave local artists a formal role in its regeneration programme by establishing an Artist Enterprise Zone in Barking town centre, along with a pilot housing scheme. Numerous apartments are provided at a discounted rate to people on low incomes working in creative industries, who are in turn required to help with community-based arts programmes. In a newspaper interview, the strategic director of growth for Barking and Dagenham Council said: 'One of the issues for us is how do we create that sense of place so we don't just build lots of housing. It seems to us that artists and the creative industries are fundamental to that.'

③ Tackling urban environmental stress

▶ *Why are environmental, transport and housing strategies important for the place-remaking process?*

Creating liveable places

Chapter 1 explored the importance of physical site conditions in determining how and why places first develop historically. In turn, human occupation modifies places in a multitude of physical ways over time, including: vegetation clearance and the production of impermeable surfaces; river diversions and the canalisation of waterways; the building of bridges, embankments and other relief modifications; coastal defences. Alongside these deliberate environmental interventions, economic development processes give rise over time to what economists call negative externalities: in this case, the accumulated effects of water, air and land pollution, along with traffic congestion and housing shortages.

- If an area has undergone large-scale deindustrialisation, there may a legacy of environmental blight (derelict factories and wasteland) which needs dealing with if new leisure services and investors are to be attracted.
- Communities in towns and cities which are economically healthy may nonetheless suffer from traffic and housing issues which impact negatively on people's quality of life.

This next section offers a brief overview of these important environmental issues.

Urban air quality

In the past, high levels of particulate (solid) and photo-chemical air pollution in British cities (because of heavy industry and an over-reliance on burning coal for heat) resulted in an untimely death for many people. In the UK, December 1952 is remembered for London's 'killer smog' which is estimated to have killed more than 4000 people (see Figure 5.17). Similar challenges persist in other countries today: according to the World Health Organization, 500 million Chinese people will, on average, die five years prematurely on account of the so-called 'airpocalypse' unleashed by China's recent economic advancement.

Great progress has been made in the UK in terms of improving air urban quality; the Clean Air Act of 1956 introduced tough controls and began the progressive easing of coal use that resulted recently in a day of 'zero coal' power generation (see page 94). Dirty polluting industries have moved abroad (for instance, to

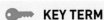

KEY TERM

Negative externalities
Costs which arise on account of economic activity, including uncompensated-for environmental or social damage.

China) while cleaner, post-industrial economic activities, such as tourism and leisure, have replaced them. This illustrates an important point about place connectivity: improved air quality for urban places in the UK is interlinked with the deteriorating conditions experienced by people in Chinese cities.

All major UK cities have their own pollution-reduction strategies, for example Park and Ride in Oxford. In 2017, the UK Government announced its intention to increase the number of urban clean air zones (CAZ) in UK cities, in the hope of tackling diesel vehicle fumes. Another action is the doubling of the congestion charge in central London for especially polluting vehicles (taking the total daily fee to more than £20).

▲ **Figure 5.17** The London smog of 1952 made some neighbourhoods in central London, including Whitechapel, life-threateningly dangerous places to live or work

Urban drainage and water quality

Urbanisation affects the water cycle: the introduction of impermeable surfaces results in the loss of drainage basin water storage areas, including vegetation and soil stores, and increased overland flow. As a result, water movement through urban catchments is usually characterised by flashy hydrographs (with a high peak discharge, low lag time and steep rising limb).

- Sustainable urban drainage systems (SUDS) are an approach to managing rainfall in urban areas that seek to replicate natural drainage systems.

- In practice, this can mean increasing the number of permeable surfaces so that surface water can drain in a more natural way and there is less reliance on sewers and drainpipes to remove water.

- For example, the Barking Riverside Park Land development in London has restored former areas of the River Thames floodplain. Rainwater can infiltrate the ground again during heavy rainfall.

River restoration and conservation in urban catchments

Actions taken to return a river channel to its natural state – including improved water quality and the reintroduction of specific features (such as meanders) and species (such as salmon) – are collectively called river restoration. This can be a vital part of the regeneration process in towns and cities whose economies now rely on tourism and leisure. Restoration work has been carried out in many places in the UK, including the River Wandle in south London (see Figure 5.18) and Salford Quays in Greater Manchester. In the latter case, work to restore water quality formed part of the much wider initiative to restore the River Mersey and Manchester Ship Canal (see pages 146–147).

▲ **Figure 5.18** Urban ramblers can walk the entire course of the River Wandle in southwest London now that it has been restored to a more natural state

ANALYSIS AND INTERPRETATION

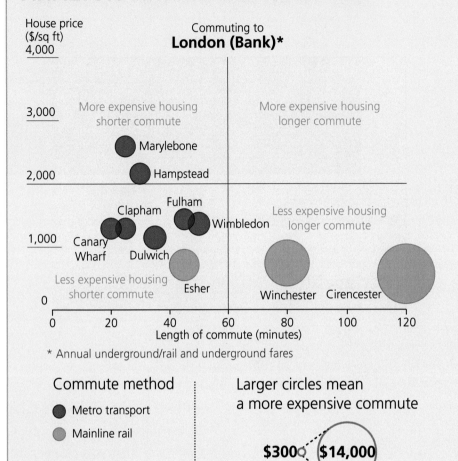

FT graphic: Paul McCallum

Sources: FT research; Knight Frank; TFL; UK operators

▲ **Figure 5.19** Commuting times and housing costs in different places

Study Figure 5.19, which shows travel times to central London and housing costs for different places located within and outside London.

(a) State the average length of travel time to central London for people from (i) Marylebone and (ii) Cirencester.

GUIDANCE

Successful completion of this task requires great accuracy. You will need to identify the centre of each proportional circle and identify the precise x-axis value that corresponds.

(b) Analyse variations in the cost of living for central London workers who live in the places shown in Figure 5.19.

GUIDANCE

In order to carry out this task successfully, you will need to do more than simply list data. It is important to also recognise the way the data have been compartmentalised into quadrants by the graph: as a result, there are three groupings of circles to mention, each of which has been categorised according to the overall cost of living factoring in both property prices and annual commuting costs. A good approach might be to provide a structured answer consisting of three short paragraphs (one for each grouping). Take care when offering supporting data to extract the information accurately. You must correctly locate the centre of each circle in order to read off the corresponding x-axis and y-axis values.

(c) Suggest ways in which the long travel times shown in Figure 5.19 could affect people's experience of living in places like Winchester and Cirencester.

GUIDANCE

This question in part relates back to the contents of Chapter 2, which explored place feelings, meanings and experiences. The resource deals with the experiences of a particular demographic group – commuters. An answer to this question could include the following or other themes.

- Spending up to four hours a day commuting from Cirencester to London and back leaves commuters with very little time to become involved with local community groups or participate in local activities other than at the weekends. This could mean that commuters do not always identify strongly with the place where they live; in turn, this might affect community cohesion in places where many people spend long hours commuting.
- The high costs of commuting might leave a typical person with less disposable income to spend on leisure and activities in the place where they live. This also impact negatively on their sense of belonging there.
- Places where large numbers of people are absent for many hours of the day because of long commuting distances can gain a reputation as 'dormitory' settlements that lack vibrancy. This might lead to those places being represented negatively by the media.

Transport and housing issues

For many people in the UK, transport and housing issues will be the determining influence on how 'liveable' the place where they inhabit actually feels. Two-thirds of the national population (around 40 million people) are of working age and the majority travel from their home to a place of employment every weekday. For many people, choosing where to live is determined in large part by the liveability balance which must be struck between housing costs (determined largely by availability and demand) and commuting costs.

- Since the late 1990s, the most pressing urban stress issue for many towns and cities, especially those in southeast England, has been the widespread lack of affordable housing, particularly for key workers.

- Young people already burdened with student debt must fork out many hundreds of pounds each month to rent a bedroom. Many can see no way forward for ever owning a home of their own in most London boroughs.
- New social divisions are therefore growing in parts of the UK. Young adults from wealthier families can rely on 'the bank of mum and dad' to help them buy a home and put down permanent roots in places such as Clapham or Wimbledon in London; but those from poorer families lack the same advantages. As a result, the long-term cohesion and sustainability of communities in these areas is threatened by a wealth dividing-line which has left a large proportion of younger people unable to make a long-term housing investment in their home place.

For those who cannot borrow from their parents, one solution is to remain at home until late into their twenties or thirties – but this can put pressure on family relationships. Young adults who want a home of their own may feel the only option is instead to move further out into the commuter belt where house prices are usually cheaper. Yet mounting pressure on the railways has left many commuters on popular lines unable to find a seat when travelling to work, despite having paid a high price for their travel. As a result, many long-distance commuters may feel theirs is an unsustainable lifestyle. Unfortunately, no effective political solution for the UK's affordable urban housing shortage is currently on the horizon.

Rough sleeping and defensive architecture

One final management challenge is the recent increase in UK homeless numbers. By one estimate, one-in-two hundred citizens were homeless in 2018. The main causes of this crisis are explained elsewhere in this book and include: job losses linked with austerity measures; the global financial crisis; technological changes affecting workplaces; and unaffordable housing in many places, especially southern England. Most of those who are counted as homeless are not rough sleepers and move from place to place staying with family or friends – so-called 'sofa surfers'. Many use hostels. But up to 8000 are rough sleepers at grave risk of suffering from hypothermia, mental or physical health problems, assault or sexual exploitation.

Certain local places attract disproportionately high numbers of rough sleepers, including central areas of Brighton, Exeter and Cardiff, along with many parts of London, including Windsor. If you judge a society by the way its most vulnerable members are treated, then some recent 'zero tolerance' measures to deal with rough sleeping may give you pause for thought.

- Rough sleepers have been effectively banned from some major railway stations, including Cardiff Central and London Victoria, under the Withdrawal of Implied Permissions (WIP) scheme.

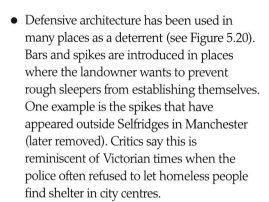

- Defensive architecture has been used in many places as a deterrent (see Figure 5.20). Bars and spikes are introduced in places where the landowner wants to prevent rough sleepers from establishing themselves. One example is the spikes that have appeared outside Selfridges in Manchester (later removed). Critics say this is reminiscent of Victorian times when the police often refused to let homeless people find shelter in city centres.

In contrast, campaigning groups like Shelter and the Joseph Rowntree Foundation continue to offer support to rough sleepers. Other charities campaign tirelessly to draw attention to the underlying issues, including lack of affordable housing and insufficient community support for vulnerable people at risk of becoming homeless. Technology is being used to help rough sleepers too: the StreetLink phone app lets local citizens contact charity workers if they find someone sleeping rough in their neighbourhood who needs assistance.

▲ **Figure 5.20** Defensive architecture is designed to deter rough sleepers from settling in particular areas. Some people view it as a necessary measure to improve the appearance, experience and 'liveability' of these places for everyone else; others perceive social injustice against vulnerable people

 Evaluating the issue

▶ *To what extent have actions to create sustainable places been successful?*

Identifying possible contexts and criteria for judging success

The focus of this chapter's debate are actions to create more sustainable places and the extent to which their goals were met successfully.

The first thing to note when setting the scene for debating this issue is the multi-strand character of sustainability (and sustainable development). However, for somewhere to be sustainable – and for it to be a pleasant and 'liveable' place for all

its citizens – interlinked social, economic and environmental goals must be met. Possible questions to ask when investigating the sustainability of a place are:

- What are the economic policies for job creation?
- Are there social policies for community cohesion and welfare?
- Do environmental policies exist for improving air and water quality?
- How are housing and transport issues being dealt with?

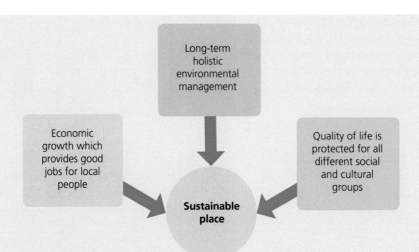

▲ **Figure 5.21** A sustainability and liveability road map for places

The concept of sustainability refers to development that meets the needs of the present without compromising the ability of *future generations* to meet their own needs. In order to debate the issue properly, we also need to think critically from the start about how best to judge whether any action has been 'successful' or not. Table 5.3 sets out some important criteria for success.

Evaluation criteria	Issues to consider
How do the perspectives of local people differ?	The place-remaking process is often contested, as many examples in this book show. This is because every stakeholder in a place may see things differently. ■ If a flagship redevelopment displaces large numbers of people from their homes, they may all subsequently judge the strategy 'a failure'. ■ Some people may judge the destruction of a particular landmark by a new development as so grievous that it cancels out any gains. ■ The 'success' of gentrification does not always trickle down to lower-income groups.
Can the success of a strategy be 'benchmarked'?	'Benchmarking' means setting a goal in advance against which progress can later be measured, such as a target number of new visitors. Like meeting the grade in an exam, the judgement of 'pass' or 'fail' can be made. Quantitative benchmarking measures might include an employment goal or a target percentage reduction in the number of vacant properties in a town centre. People can be asked to report how happy they feel (this is their 'lived experience' of a place). The Mappiness project is an example of how this can be done (see page 73).
Is there a wider geographic context to consider?	Benchmarking is not always as simple as it sounds. This is because the success achieved by a local strategy or action is also dependent on the wider regional, national and global contexts in which every local place is embedded. ■ If a country experiences an economic recession, a local regeneration strategy may achieve less than was hoped for. But should the strategy itself then be viewed as a failure? ■ To what extent do the economic benefits of city centre regeneration trickle down in measurable ways into poorer surrounding neighbourhoods and communities?

▲ **Table 5.3** Ways of judging whether actions and strategies have been successful

Pearson Edexcel

AQA

OCR

WJEC/Eduqas

Evaluating the view that actions to create sustainable places have succeeded

In support of the statement, we return to the work of the Mersey Basin Campaign (MBC) previously featured in Chapter 4 (see page 146). The MBC worked alongside public and private stakeholders during the 1990s to restore water quality throughout the River Mersey basin, including the Manchester Ship Canal. Figure 5.22 shows the formal strategic sustainability plan adopted by the MBC at the very outset. As you can see, they certainly set out to tick all the sustainability boxes. The restoration of Salford Quays, a place located to the west of Manchester city centre, is widely regarded as a success story of place remaking that simultaneously improved local social, economic and environmental conditions.

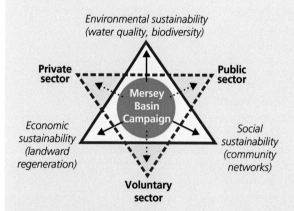

▲ **Figure 5.22** The Mersey Basin Campaign successfully used a sustainability framework to bring positive environmental, economic and community changes to waterside places in the northwest of England

Salford Quays is an area on the western edge of Manchester where a series of very large docks join the Manchester Ship Canal (which, in turn, links with the River Mersey).

- Intense industrial activity in Salford Quays left the water heavily polluted. By the time the old dock industries were closing down because of

global shift in the 1970s, the filthy water was a major disincentive to anyone seeking to invest in dockside redevelopment.

- The worst-affected area was the Turning Basin – 28 hectares of foul-smelling and bubbling open water (caused by escaping methane and hydrogen sulphide). Oxygen levels were completely depleted by rafts of decomposing human sewage (a product of poorly-maintained sewers).

- Salford Quays benefited greatly from the improvements in water quality made to the Manchester Ship Canal and River Mersey – funded largely by water company United Utilities – that were outlined in Chapter 4 (see page 147). At Salford Quays, uniquely-designed oxygenation injectors were installed in 2001 at a cost of £4.5 million (see Figure 5.23). Up to 15 tonnes of liquid oxygen can be added every day to help restore and maintain water quality.

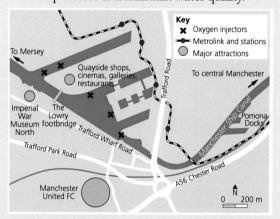

▲ **Figure 5.23** Without water quality improvements at Salford Quays, it is unlikely that the bankside developments shown here would ever have found funding

Beyond resolving the environmental problems, this action was the first step towards economic and social regeneration of the area. As water quality improved, new housing and commercial developments funded by private-sector players such as Peel Holdings sprung up around the water's edges.

- Salford also benefited from media exposure when the Triathlon World Cup was staged there in 2003. Photographs of swimmers in the previously polluted water may have helped change outsiders' perceptions of the area: well-educated young professionals have since settled in Salford in large numbers.

- The district now supports 13,000 jobs and sees around 4 million visitors a year, who find themselves in a clean, attractive setting where they can enjoy significant flagship developments such as the Lowry and the Imperial War Museum North. Salford was chosen by the BBC as its new production site in 2004 (see Figure 5.24).
- When *The Financial Times* newspaper ranked Greater Manchester as the UK's most socially 'vibrant' urban area in 2015 (based on combined census and commercial data over a ten-year period), it singled out Salford as a particularly successful area.

▲ **Figure 5.24** Salford Quays today is viewed by many people as an environmentally, economically and socially sustainable place

Looking back at Figure 5.22, all of the Mersey Basin Campaign's sustainability goals appear to have been met. Yet not everyone agrees the regeneration of Salford has been an *unqualified* success: some schools around the edges of the area serving long-established parts of the local community are currently underperforming. A University of Salford report recently showed

that ten per cent of the area's rented housing was below a decent standard and Labour-run Salford Council remains concerned that some social tenants are living in 'slum-like' conditions. This suggests that economic prosperity has not yet trickled down to any great extent into many of the perimeter and neighbouring areas.

Evaluating the view that actions to create sustainable places have failed

The counter-argument is that it is nigh-on impossible to create truly sustainable places because of a systemic failure by UK policy makers to confront several pressing issues. These are:

- a current social crisis of housing supply and affordability in many places in the UK
- medium-term economic threats, including AI and robotics (see page 95)
- the long-term local and global environmental implications of urban ecological footprints.

Of these, housing issues are the greatest immediate concern. Arguably, the sustainability of many place-remaking schemes throughout the UK is jeopardised already by the acute lack of access younger people now have to housing in the places they call home. This is especially true of central urban areas and popular rural market towns where housing demand greatly outstrips supply (see page 172). Since 2010, the number of homeowners aged under 45 in England has fallen from 4.5 to 3.5 million (see Figure 5.25). Many young people are forced to live in their parents' home as young adults or to rent properties whose fees are so high they find it impossible to save enough money to afford the deposit needed to buy a home of their own.

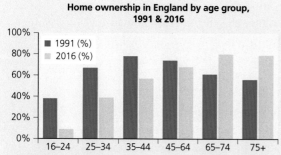

▲ **Figure 5.25** The changing demographic pattern of home ownership in the UK has worrying implications for the social sustainability of local communities throughout the country

In 1997, the median house price in the UK was roughly three times median annual income. Now it is almost eight times higher than typical earnings. For decades, successive governments have failed to build enough homes to accommodate a growing population despite continued warnings from experts that demand has vastly outstripped supply because of:

- an ageing population (today's longer-lived elderly occupy housing far longer than previous generations did)
- rising affluence, which allows more families to own a second home
- planning laws preventing new housing being built in protected countryside
- immigration into the UK, particularly from eastern Europe
- **property asset purchasing** by offshore players looking for safe ways to invest money
- more young people leaving home aged 18 to attend university, thereby creating housing shortages in university towns and cities.

Experts say that to keep pace with demand, up to 300,000 new homes should be built each year, but this has not happened since the 1970s and around half that number are actually completed each year. NIMBY attitudes – where local people use planning laws to block new housing developments in their own home place (see page 18) – sometimes help explain the slow rate of progress.

Given there is a national housing shortage, successful place remaking – which attracts incomers to an area, thereby forcing local land prices even further upwards – is invariably viewed as a 'mixed blessing' by long-term residents. For those who own a home, rising prices can amount to a 'lottery win' should they choose to sell and move on. For those in social housing or who rent privately, it is a very different story.

- London's Haringey Council has plans to demolish the Northumberland Park and Broadwater Farm estates as part of a redevelopment plan to build 6400 new homes. Although the plans incorporate green space and may appear environmentally sustainable, many residents are worried about where they will be rehoused once the redevelopment is complete. They are naturally sceptical about how much affordable housing will be made available, particularly given the squeeze on local council budgets since austerity measures began in 2010.
- Page 111 examined the 2017 fire that destroyed 24 storeys of social housing at Grenfell Tower in gentrified North Kensington. There is grim irony in the fact that almost ten years earlier Kensington and Chelsea Council had launched a decade-long community strategy that made 'developing a sustainable community' its top priority.
- In Streatham in southwest London, a one-bedroom flat typically cost one-third of a million pounds in 2017. It is hard to imagine many of the

---

🔑 **KEY TERM**

Property asset purchasing Homes in a local place are bought as investments by outsiders rather than local people who need them. After the GFC, many foreign investors regarded London properties as relatively safe assets to buy into. As a result, up to one-in-three newbuild homes in prime London locations are empty as part of the 'buy to leave' phenomenon (the owner does not want to live there and is merely content to own the property and watch it rise in value).

children currently attending primary school in Streatham will be able to afford a property of their own once they reach working age. This threatens the sustainability of the local community because there is a clear barrier against the next generation ever being able to own a home in the place where they have grown up.

There are requirements in the UK that new property developers provide a percentage of homes in new projects at affordable prices. However, the percentage is often very low: for example, developers in central Leeds only need meet the requirement that 5 per cent of homes in each new project are made available at affordable prices. This is defined as 80 per cent of the market rate – which still represents a great deal of money to many people.

Arriving at an evidenced conclusion

To what extent have recent actions to create sustainable places actually succeeded? Views about the success or otherwise of any strategy or initiative will, of course, in all likelihood diverge. This is because different stakeholders – including local and national governments, local businesses and residents – all assess success using contrasting criteria. Governments may value the 'big picture' of job creation while local people's views depend more on their own lived experiences of the changing places they occupy.

Economically-sustainable places such as London's Isle of Dogs have continued to successfully attract new rounds of investment through the operation of cumulative causation

('snowballing') processes. Yet, as we have seen in this chapter, economic overheating, rising house prices and commuting issues can all conspire to make economically-successful locations feel less attractive and liveable places to be for some of their residents.

Not everyone in many of the UK's gentrified inner cities would agree these places have anything approaching a sustainable future given the utter unaffordability of housing for ordinary people in a growing number of areas. Even Salford Quays – where a properly holistic vision of place remaking has yielded truly transformative environmental, social and economic effects – is not immune to the criticism that some local people remain socially excluded from local housing and employment markets.

Finally, it is important to remember that the success and sustainability of any local place is, to a large extent, also determined by the underlying economic fortunes of the larger-scale cities, regions and countries within which they are embedded. In the case of local places in the UK, their economies have been buffeted first by the global financial crisis and more recently by the uncertainty which lies ahead on account of the country's changing relationship with the EU. In the event of large numbers of European migrant workers quitting London and other major cities, there are implications for businesses and housing markets in many localities. It remains to be seen whether the long-term knock-on effects of the Brexit vote hinder or enhance the longer-term prospects for creating sustainable places in the UK.

Chapter summary

✔ Local and national government play a very important role in driving economic change using policies such as enterprise zones. Policies to stimulate economic development are grounded in economic theories and models that have merits but can also be critiqued.

✔ Enterprise zones were first used in the 1980s and were recently revived after the global financial crisis. Perspectives vary on the success of enterprise zones in general; however, there is wide agreement that the London Docklands Development Corporation carried out exemplary work in redeveloping the Isle of Dogs as a financial services hub.

✔ Gentrification is a controversial economic and social process with many downsides accompanying its upsides. Housing costs and shortages in gentrified places are a wicked problem lacking any easy solutions.

✔ During the past 40 years, British society has in general become more cohesive and inclusive. Great efforts have been made to tackle the causes of cultural tension in the UK and to manage diversity successfully. A range of (local and national) government policies and new institutional strategies have sought to foster community cohesion with some successful results.

✔ Although urban places are often far more liveable than they were in much of the twentieth century, it remains the case that poor air quality, transport problems and housing shortages have created a range of persisting urban stresses.

✔ Strategies to create more sustainable places have often had mixed results. Great steps have been taken to improve the overall liveability of some places; Salford Quays is a particular success story.

✔ Housing supply issues throughout the UK threaten the future of many places though, and may jeopardise the longer-term success and sustainability of the place-remaking schemes featured in this and other chapters.

Refresher questions

1 What is meant by the following geographical terms? Growth pole; cumulative causation; trickle-down; enterprise zone.

2 Outline the incentives for new industries created by enterprise zones.

3 Explain functional, demographic and environmental changes affecting the Isle of Dogs since the 1980s.

4 Using examples, explain the causes of cultural tension in British and American cities.

5 Compare the costs and benefits associated with the gentrification of inner-urban areas.

6 Using examples, suggest ways of tackling cultural and social tensions in urban areas.

7 What is meant by the following geographical terms? Overheating; decanting; gated community; negative externalities.

8 Using examples, explain ways of making urban places more environmentally sustainable.

9 Explain why many people view housing as the most pressing challenge for many urban places.

Discussion activities

1 In groups, discuss the view that more money should be raised through national taxes to fund economic regeneration strategies in local places that are in greatest need of assistance.

2 In groups, compare the cumulative causation model with the classical economic theory that argues that prosperous places will lose their initial advantage over time (because of rising labour and land costs). What are the strengths and weaknesses of the arguments these models and theories use?

3 Discuss the view that regeneration and gentrification processes inevitably create 'winners' and 'losers'. Devise a checklist of possible steps a local authority might take in order to limit the negative impacts of gentrification for some groups of people.

4 In pairs, design a mind map showing how different cultural and ethnic groups have been included in various categories of media text produced in the UK (television shows, films, advertisement) that you are familiar with. Why are inclusive representations viewed as important?

5 In groups, discuss your own attitudes towards the commuting lifestyle that many people in the UK have adopted. If you find yourself working in a city centre one day, would you prefer to live expensively close to where you work or more cheaply at a greater distance away?

6 Discuss what needs to be done about the UK's housing crisis in order to help people currently aged 16–21 to get on to the 'housing ladder' sooner once they start working. Should more housing be built and, if so, where? Should people who rent houses be given the right to buy them? Should laws be passed to stop asset property purchasing? What else could be done?

FIELDWORK FOCUS

The topics covered in this chapter provide many opportunities for an A-level independent investigation exploring the economic, social or environmental impacts and issues of place remaking. For AQA students, urban environmental and sustainability themes form an important part of their A-level course and can also provide a topic focus for the independent investigation.

A *Using a mixture of questionnaire data and secondary sources to investigate the gentrification of a place.* A carefully-selected neighbourhood can be chosen for investigation which, prior research shows, has undergone or is undergoing gentrification. Secondary data could include: house price changes over time; the socioeconomic profile of the place in the 2001 and 2011 Censuses; media representations of the place (some gentrified places will have featured in newspapers, for example 'the top ten cool places to live'). Opportunities for primary data collection could include: interviews with users of local services and transport links; photographs of new fashionable bars, restaurants and businesses. The investigation might focus on the scale of any changes which have occurred or on diverging viewpoints about the desirability of these changes.

B *Interviewing people in order to find out more about commuting patterns and the decision-making process affecting where people live.* A well-chosen sample of commuters could be interviewed in order to find out more about the typical distance people are prepared to travel to work in a particular place. You will need to think carefully about your methodology, however. When people are rushing to work or to catch their train home, it could be a bad idea to try to stop them and ask questions. A better idea might be to conduct interviews at lunchtime in a street full of cafés that local workers use in large numbers.

C *Investigating an important liveability issue in your home place.* There are many opportunities for carrying out an interesting or unusual independent investigation dealing with a particular social or environmental issue. Homelessness, rough sleeping, air pollution and waste management are all relevant issues that could form a legitimate focus for your independent investigation. However, before attempting anything like this you must think carefully about the practicality of carrying out any work, along with safety and ethical issues. You would need to undertake a proper risk assessment before carrying out primary data collection. There may be useful smartphone apps that can be used to measure, for example, air quality.

Further reading

Amin, A. (2012) *Land of Strangers.* Cambridge: Polity.

Andreou, A. (2015) Anti-homeless spikes: 'Sleeping rough opened my eyes to the city's barbed cruelty'. *The Guardian* [online]. Available at: www.theguardian.com/society/2015/feb/18/defensive-architecture-keeps-poverty-undeen-and-makes-us-more-hostile [Accessed 21 February 2018].

Barczewski, S. (2000) *Myth and National Identity in Nineteenth-Century Britain: The Legends of King Arthur and Robin Hood.* Oxford: Oxford University Press.

CerealKillerCafé.com. Available at: http://www.cerealkillercafe.co.uk.

Florida, R. (2017) *The New Urban Crisis: Gentrification, Housing Bubbles, Growing Inequality and What We Can Do About It.* London: Oneworld.

Gilroy, P. (1986) *There Ain't No Black in the Union Jack: The Cultural Politics of Race and Nation.* London: Hutchinson.

Khomami, N. and Halliday, J. (2015) Shoreditch Cereal Killer Cafe targeted in anti-gentrification protests. *The Guardian* [online]. Available at: http://www.theguardian.com/uk-news/2015/sep/27/shoreditch-cereal-cafe-targeted-by-anti-gentrificationprotesters [Accessed: 23 March 2018].

Larkin, K. and Wilcox, Z. (2011) What would Maggie do? *Centre for Cities* [online]. Available at: www.centreforcities.org/wp-content/uploads/2014/09/11-02-28-What-would-Maggie-do-Enterprise-Zones.pdf [Accessed 21 February 2018].

Lees, L. Slater, T. and Wyly, E. (2010) *The Gentrification Reader.* Abingdon: Routledge.

London Borough of Barking and Dagenham (2016) *Barking Artists Enterprise Zone* [online]. Available at: www.lbbd.gov.uk/wp-content/uploads/2016/04/Barking-Artist-Enterprise-Zone.pdf [Accessed 21 February 2018].

Massey, D. (2007) *World City.* Cambridge: Polity.

May, J. (2014) Exclusion. In: P. Cloke, P. Crang and M. Goodwin. *Introducing Human Geographies.* Abingdon: Routledge.

Myrdal, G. (1957) *Economic Theory and Under-Developed Regions.* London: Duckworth.

Oakes, S. (2017) Wicked problems. *Geography Review,* 30(4), 25–27.

Whitehead, M. (2006) *Spaces of Sustainability: Geographical Perspectives on the Sustainable Society.* Abingdon: Routledge.

CHAPTER 6

Issues for rural places

Since the 1970s, external forces have been reshaping different types of rural places in diverse ways. These changes have sometimes given rise to new kinds of rural identity but they have been a cause of tension and conflict, using a range of examples of rural places and communities too. This chapter:

- explores how rural places have been shaped by past and present connections with other places
- investigates economic and demographic changes and challenges for diverse types of rural places
- analyses the extent to which different attempts at rural place remaking have been successful
- discusses how far social attitudes towards rural identity vary among different groups of people.

KEY CONCEPTS

Differentiated countryside The idea that any analysis of the countryside will reveal varying types of rural places with very different economic, social, political and environmental identities. These different identities – or 'ruralities' – often arise because of how far from urban areas a rural place is located.

Wilderness Remote areas whose unspoilt characteristics have ecological, scientific and also cultural and aesthetic value. The scale of wilderness can vary; large continental areas of the Americas are defined as wilderness, but so too are small corners of rural Britain, such as the granite tors of Dartmoor or remoter parts of the Pennines. The concept is contested insofar as many so-called wilderness places have been modified by humans; landscapes which we think of as natural, such as the Lake District, were extensively changed in the past (when forest was cleared for sheep grazing).

 # Rural places, players and connections

▶ *In what ways have rural places been shaped by past and present connections with other places?*

Shifting flows of people, investment and resources

Previous chapters focused mainly on how urban places have been affected by exogenous forces such as global shift, new technologies and international migration. Rural places experience similar forces of change. They too are continually reshaped by shifting flows of people, resources and investment at

"I hope this client knows what he's doing — giving us a bob for every pylon we knock down on his farm."

London Express Service

▲ **Figure 6.1** Rural places were transformed by central government's roll-out of the national grid in the 1950s and 1960s. This newspaper cartoon from 1965 provides qualitative evidence of the controversy caused by the reshaping of local places by exogenous political forces (the artist, Carl Giles, was vexed by the fact that pylons had recently been installed near his own house)

varying scales, with a range of demographic, socioeconomic and cultural impacts.

- The functional transformation of large parts of the British of countryside into 'post-productive' places was previously explored in Chapter 1 (pages 11 and 23).
- Many rural places have been transformed by population movements from neighbouring urban areas; demand for homes among affluent **counterurban migrants** explains the widespread lack of affordable housing in many popular villages. **Rural gentrification** results in the gradual displacement of lower-income families.
- National-scale forces and players, including the UK Government and large companies such as Sainsbury's, exert influence over rural places. Strategic infrastructure decisions by central Government can be a 'game changer' for rural areas (see Figure 6.1). Present-day controversies include the selection of sites for new airport runways (Heathrow), railway routes (High Speed 2) and power stations (Hinkley Point C). In the past, rural villages were sacrificed to reservoir construction (see Figure 6.2). Governments use compulsory purchase orders as a

▲ **Figure 6.2** The drowned village of Derwent in Derbyshire's Ladybower Reservoir. The church tower slowly disappeared below the water as the reservoir filled in 1946

🔑 **KEY TERMS**

Counterurban migrants People who have relocated from cities to rural towns and villages. Some are working-age people, often with young families; others are retirees.

Rural gentrification Mirroring the process of urban gentrification, this involves new middle-class counterurban migrants renovating old housing stock in rural towns and villages, eventually driving prices beyond the reach of ordinary longer-established families.

planning tool in cases such as these, leaving rural landowners with little redress. The influence of companies is demonstrated by the arrival of new private-sector retail parks in a rural setting, such as Bluewater in Kent. These new developments create employment but bring traffic congestion and pollution too.

● Global-scale players impacting on rural places include TNCs with a stake in retail parks and sovereign wealth fund investors in railways, power stations and other infrastructure projects (see pages 128–129).

CONTEMPORARY CASE STUDY: THE SHIFTING SANDS AND FLOWS OF FORMBY'S EDGELANDS

In their book *Edgelands*, Paul Farley and Michael Roberts analyse the characteristics and meanings of places that are neither urban nor rural. Parts of the Sefton Coast, north of Liverpool, fit this description well. They are the strips and pockets of land immediately adjacent to densely-populated areas within Liverpool's much larger rural–urban fringe, including expanses of dunes. Sefton has the largest dune area in England, extending over 17 km and up to 4 km wide in some locations (see Figure 6.3). Because of their proximity to housing in Crosby, Formby, Ainsdale and Southport, these dunes can be categorised – in the view of local author Jean Sprackland – as edgelands.

The dunes are highly mobile and their shifting sands have swallowed up housing in the past. Secondary data sources in local libraries tell of fifteen houses enveloped by dunes following a storm in 1670. Over time, this edgelands environment has been used in a multitude of ways. Today, the dunes and beach between Southport and Crosby are a palimpsest (see page 32) of golf courses, rusting industrial structures, sports pitches, burned-out cars, abandoned wartime shelters, drainage ditches, a military rifle range, one small airport, railway sidings, National Trust woodlands, plastic beach debris, grazing sheep, the *Star of Hope* shipwreck and 100 cast-iron, seaward-gazing human statues sunk in Crosby's intertidal zone by sculptor Antony Gormley.

Both in the past and present, this has always been a highly-connected place. Historically, Formby's coastal edgelands have been linked with other places at regional, national and global scales by flows of people, sand, food and waste. Many different geographies and stories intersect in this place.

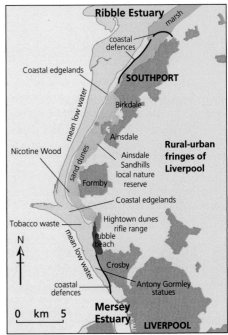

▲ **Figure 6.3** The strip of coastal land between Southport and Liverpool is a dynamic area of edgelands (set within Liverpool's wider rural–urban fringe) that has been reshaped many times by physical and human forces

■ Large areas of the inland dunes were developed for housing in the 1960s. Formby's population exploded from 5000 to about 30,000 around the same time as new towns like Milton Keynes were being built.

■ At one time, the dunes were a nationally renowned site of asparagus production. Today, small-scale, organically produced Formby asparagus is sold at London's Harrods and

Covent Garden market. Before Liverpool's sewerage system was completed in the late 1800s, large volumes of 'night soil' (human excrement) were brought daily to Formby by railway for use as an asparagus fertiliser.

■ Large volumes of fine windblown sand were excavated from the dunes between the 1930s and 1950s for use by the region's glass-making industries, including Pilkington in St Helens. The sand was also perfect for drop-forging firms in distant places like Birmingham and Coventry.

■ Vast amounts of tobacco waste were dumped in the dunes between 1956 and 1974. They came from Liverpool's tobacco industry; in turn, the tobacco itself originated in China, Brazil, Cuba, Poland, India and the USA. The presence of tobacco waste had everything to do with the port of Liverpool's past role as a major global industrial hub connected by flows of resources and people with a multitude of distant places. Today, the tobacco industry has gone and Liverpool's enormous Stanley Dock Tobacco Warehouse lies empty with broken windows (although there are plans to redevelop it into luxury housing). Lingering memories of the tobacco trade and its waste are found in Formby too, with local place names like Nicotine Path and Nicotine Wood.

In January 2014, the Sefton coast experienced the biggest storm surge since 1953. A 9.8 m tide combined with a severe westerly gale and low atmospheric pressure to generate exceptionally high water. Large destructive waves attacked the dune frontage, causing great damage especially at Formby Point. Figure 6.4 shows huge chunks of tobacco waste spilled from broken dunes on to the beach below. As a useful exercise, sketch out a mind map that explains all the events which led up to this happening. You will find yourself linking together a range of different (past and present) human and physical processes and flows, all of which converge at Formby's coastal edgelands.

▲ **Figure 6.4** Dune erosion by a storm in 2014 revealed waste from the global tobacco trade dumped in Formby's edgelands during the 1950s

The role of external players in driving rural change

Many groups and organisations help reshape rural places over time through their collective governance. This term describes a more diffuse power structure than government and includes the actions of many players. Some hold direct power because they have unrestricted access to the funding places need, while others exercise a broader influencing or advisory role. Complex actor networks may often be involved in the governance of rural areas that politicians regard as problem places, such as the rural coast and islands of Strathclyde, Scotland (see Figure 6.5).

Where a rural community's future is at risk because of economic, demographic or technological changes, place-remaking actions become necessary in which both top-down and bottom-up players have a role. However, conflict may ensue if some stakeholders wish to preserve the landscape while others hope to exploit it economically. Typically, there may be conflict between productive activities (farming, which continues to dominate rural land use in many regions), consumption interests (leisure and tourism industries) and environmental movements (wishing to see no development at all). These tensions are explored further on page 200.

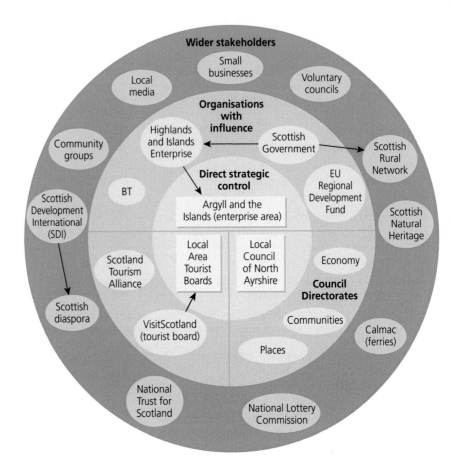

Wider stakeholders

Small businesses

Local media

Voluntary councils

Community groups

Organisations with influence

Highlands and Islands Enterprise

Scottish Government

Scottish Rural Network

Direct strategic control

BT

Scottish Development International (SDI)

EU Regional Development Fund

Argyll and the Islands (enterprise area)

Scottish Natural Heritage

Scotland Tourism Alliance

Local Area Tourist Boards

Local Council of North Ayrshire

Economy

Council Directorates

Scottish diaspora

VisitScotland (tourist board)

Communities

Calmac (ferries)

Places

National Trust for Scotland

National Lottery Commission

▶ **Figure 6.5** Governance of the rural coasts and inner islands of western Scotland involves a complex actor network composed of both internal (endogenous) and external (exogenous) players

CONTEMPORARY CASE STUDY: THE GLASTONBURY FESTIVAL

This case study provides a brief analysis of the external players and connections that have shaped the Glastonbury Festival in Wiltshire into one of the world's most successful music events. Short-lived festivals (focused on music, arts, food, sports or other interests) are interesting case studies of place

reimaging because they are ephemeral (temporary). Glastonbury is the largest of around 20 major music festivals staged outdoors in the UK. Almost 200,000 people attend the 72-hour event, which is held in a scenic rural landscape close to Glastonbury Tor (see Figure 6.6). The festival site itself does not have any

▲ **Figure 6.6** For five days of the year, the Glastonbury Festival site near the village of Pilton temporarily becomes a 'pop-up city' and the UK's 46th largest settlement by population size

special ecological significance and is used as pasture for cattle grazing during the rest of the year. The grass has remarkable resilience and can recover from the 'scorched earth' trampling effect that hundreds of thousands of dancing feet bring over the festival period.

However, great stress is placed on local residents for the duration of the festival, including noise and nuisance from all-night revellers. This inevitably brings opposition. Glastonbury promoter Michael Eavis has fought numerous legal battles with his neighbours since the festival was first held in 1970.

Staging the event now every two years, rather than yearly, is one compromise that has been reached with those parts of the community who remain opposed to the festival. Other local people like the business it brings. Hotels and guesthouses can charge exorbitant prices to celebrities and senior music industry managers. A significant proportion of local small-business annual earnings can be generated during the five-day festival period.

Figure 6.7 shows how the Glastonbury site becomes a globally-connected place for the duration of the festival. Flows of people, money and information operate at scales ranging from local to international. This temporary flow network is co-constructed by a range of players.

- Festival-goers come from the local area, the UK as a whole and other countries around the world (many international visitors attend the festival; some view it as 'the trip of a lifetime' or a 'bucket list' event).

- Musicians performing at the festival come from the USA and many other countries.

- Many journalists, filmmakers and television producers will be present, including a large BBC team who broadcast the event to national and global audiences.

- Local catering companies provide food and drink on site.

In addition to major festivals such as Glastonbury, many smaller-scale festivals play a vital role supporting the year-to-year economic fortunes of some rural places. The Green Man Festival in Brecon near Abergavenny has helped put this relatively remote part of Wales on the map. A permanent association with a popular and fashionable annual festival can have an important and lasting effect on how a place is perceived by outsiders – even though each festival event does not last for long.

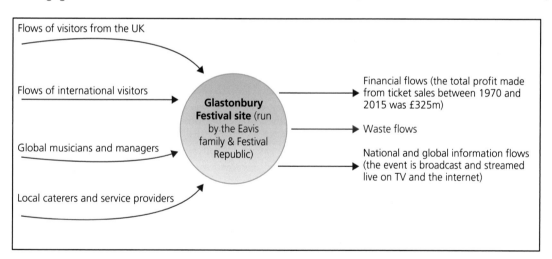

▲ **Figure 6.7** The Glastonbury Festival is a form of temporary place remaking that occurs every two years. Key local players, the Eavis family, have worked with UK company Festival Republic to generate global-scale flows of investment and people

② Change and challenge in the differentiated countryside

▶ *How have economic and demographic pressures affected different types of rural place?*

Diverse rural places

Chapter 1 explored definitions of the word 'rural' with reference to economic functions, population density and land-use criteria. In total, around 95 per cent of the UK's land area has not been built up and could therefore be considered rural (just 0.1 per cent of land is 'continuous urban fabric' under the EU's Corine Land Cover system and 4.4 per cent is 'discontinuous urban fabric'). These are interesting statistics because many people perceive the UK as being much more urban than it really is.

This rural mass is far from uniform. Chapter 1 introduced 'transitional' concepts such as the rural–urban continuum, the rural–urban fringe and edgelands. But there is also great physical, economic and social diversity among remoter rural places. During the 1990s, a team of rural geographers led by Terry Marsden identified four typical rural landscapes in the UK. These are shown in Figure 6.8. It is a taxonomy which draws on place meanings and representations as much as it does on differences in economic function and demographic character. This model of the differentiated countryside is also concerned with who holds power over these different rural places.

Economic change in the post-productive countryside

The productivist years of the British countryside lasted until the late 1970s. Since then, the dominance of agriculture has progressively been eroded by a combination of external forces, including:

> **KEY TERM**
>
> **Global agribusiness** A transnational farming and/or food production company. This blanket term covers various types of TNC specialising in food, seed and fertiliser production, as well as farm machinery, agrichemical production and food distribution.

- the growth of global agribusiness (and cheaper food imports) and the continued mechanisation of farming, resulting in far fewer work opportunities for long-established communities in places where agriculture has traditionally been a major source of employment
- counterurbanisation and the arrival in some rural places of wealthy incomers with their own views and values (often informed by the powerful place meaning of the rural idyll, which Chapter 2 explored; see page 63)
- changing social attitudes and the growth of environmental awareness, including opposition (particularly among city dwellers) to intensive farming practices and foxhunting (see page 68)
- the arrival of broadband internet services (encouraging more people to relocate to rural areas where they can work from home, while also helping rural and tourist industries to advertise their services globally).

Preserved countryside

Typical of the English Lowlands, including large areas of Surrey, Kent, Buckinghamshire and more accessible upland areas such as the Lake District, this type of rural place is characterised by the preservationist attitudes of its inhabitants, many of whom are middle-class incomers. New development plans which threaten the 'rural idyll' experience (see page 63) of such areas are likely to be challenged vigorously by counterurban migrants who have colonised and gentrified villages perceived to be rich in natural and/or cultural heritage. These NIMBY ('not in my back yard') attitudes derive from the viewpoint that rural areas have historical identities and meanings that must be protected at all costs.

Contested countryside

In areas just beyond the rural and urban fringe (and the limits of the metropolitan commuting zone), farmers and landowners may still have sufficient power and influence to bring developmental changes, especially if the landscape lacks special environmental quality. However, some remoter rural areas are starting to receive larger numbers of in-migrants because of increased opportunities to work from home using broadband internet. Place conflicts can arise when these incomers want to see the countryside preserved as wilderness, while farmers instead want a productive working landscape. Large parts of Yorkshire and Devon fit this profile.

Paternalistic countryside

The term describes rural places where the 'paternal' power of old (possibly aristocratic) estate owners and large farms has gone almost entirely unchallenged. However, in the face of falling estate incomes some major landowners are actively seeking to diversify economically. New ventures may include hunting trips, with a view to achieving economically sustainable long-term Estate management. Scotland's Isle of Jura fits this profile well: the Ruantallain Estate offers affluent tourists a deer-stalking opportunity at around £400 per stag. Other remote rural areas are used increasingly for renewable power generation.

Clientelistic countryside

In some of the UK's most remote rural areas, economic activity is not viable without state support. Counterurban migration has brought fewer incomers to such places. The remoteness, inaccessibility and poor climate in some upland areas all mean tourism revenues are limited. Local politics are dominated more by concerns about employment and community sustainability than environmental protection. Local people may be highly dependent on government grants and subsidies. Many fringe areas of the Highlands and Islands of Scotland and parts of mid-Wales fit this place profile.

▲ **Figure 6.8** Four different place profiles for the UK's 'differentiated countryside'

These and other changes left the productivist countryside in disarray. Politicians and the public alike began asking questions such as: What is the purpose of rural places? Who are they for? Rural geographers including Paul Cloke have written at length about this breakdown of 'coherence' in how we think about rural places compared with the productivist era when a greater consensus existed about the function of the countryside.

Rural employment characteristics

A snapshot of the UK today, courtesy of the Department for Environment, Food and Rural Affairs (Defra), shows the following.

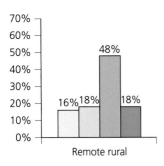

Remote rural

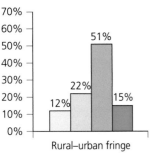

Rural–urban fringe

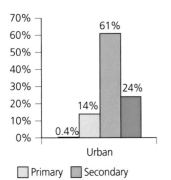

Urban

☐ Primary ☐ Secondary

☐ Tertiary – private   ☐ Tertiary – public
   sector   sector

▲ **Figure 6.9** In Scotland, a relatively high proportion of employees in rural areas are working in the primary sector. Source: Inter-Departmental Business Register, 2014

- Around 9.3 million people, or 17 per cent of England's population, lived in English rural areas in 2014.
- The size of the agricultural workforce in England and Wales has continuously declined since 1840, when it employed 22 per cent of the UK workforce (and almost all rural workers), to 1 per cent in 2011 nationally (rising to 3.5 per cent employed in agriculture, forestry and fishing in rural areas). Given that 71 per cent of land in the UK – 17.2 million hectares – is still used as farmland, this is clearly an industry that has become highly dependent on machinery.
- On average, 15 per cent of rural employment is in retail, around 13 per cent in tourism and 18 per cent in public services (government, health and education). In some places, significant numbers of people work for private employers or are self-employed.
- Figure 6.9 shows employment in Scotland as a whole. A clear rural–urban employment divide is evident. Also, there is higher dependence on primary industry in rural Scotland than in rural England.

Variations between different rural areas

Employment characteristics vary considerably *between* different kinds of rural place in the differentiated countryside.

- In the UK's most remote rural areas – typically categorised by Marsden *et al.* as 'paternalistic' or 'clientelistic' countryside (see page 191) – the proportion of working people employed in primary industries may rise to one-third.
- In some remote areas, another one-third work from home (significant numbers may be counterurban professional migrants, such as writers, artists and designers).
- In contrast, a very high proportion of people living in affluent rural places close to major cities such as London or Manchester are commuting professionals, including teachers, doctors, lawyers and bankers.
- In tourism 'honeypots', the proportion of people employed by tourist industries will be far higher (see Figure 6.10).

▲ **Figure 6.10** Employment has changed markedly in Tobermory, Scotland, in recent decades. Fewer people work in the fishing industry now but tourism thrives because the children's television programme *Balamory* was filmed in this place

Demographic changes, challenges and benefits

Fewer than half of people living in rural areas are aged below 45 years, compared with almost two-thirds in urban areas. Figure 6.11 shows a markedly higher average age for communities living in non-fringe rural areas. Once again, it is important to recognise how much these data may vary for different rural 'types'. In some remote and 'clientelistic' rural places, out-migration of the young and a lack of incomers have led to an even more 'top-heavy' population structure.

In contrast, some relatively accessible and more successfully reimaged rural locations, such as Glastonbury or Yorkshire's Hebden Bridge, have relatively youthful populations on account of relatively recent waves of counterurban migration. However, some places that attracted large number of counterurban migrants in the 1970s and 1980s have an ageing population because the incomers have now reached retirement age, for example the Isle of Arran in Scotland.

Tensions have sometimes developed in rural places most affected by counterurbanisation. In popular villages, waves of affluent incomers have driven property prices beyond the reach of traditional communities. Housing shortages worsened after the 1970s when urban residents began buying rural properties to serve as a second home or 'bolthole' from the city. In Wales, the process of rural gentrification was strongly contested in places where the cost of homes spiralled due to demand from English buyers. Between 1979 and 1981, over 200 arson attacks were carried out by Welsh nationalists on homes bought by English people as part of the Meibion Glyndŵr campaign. In this case, economic changes had become tangled up with wider issues of place identity.

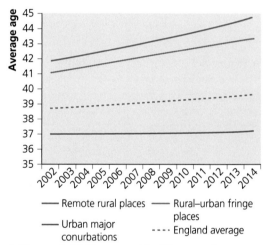

▲ **Figure 6.11** Demographic differences between rural and urban places are increasing

International migration and rural population change

Sometimes, international migrants arrive in rural areas in relatively large numbers.

- Rural farming areas surrounding Peterborough have attracted eastern European male migrant workers (though since the Brexit referendum some farms are reportedly struggling to recruit labourers).
- Hotels in rural Scotland increasingly rely on female eastern European staff.
- Holy Island in Strathclyde is home to a small population of Tibetan monks who settled there in the 1990s and have since developed a commercially-successful meditation retreat for spiritually-minded people.

In recent years, some rural geographers have investigated the experiences of traditional white working-class British people in rural places affected by

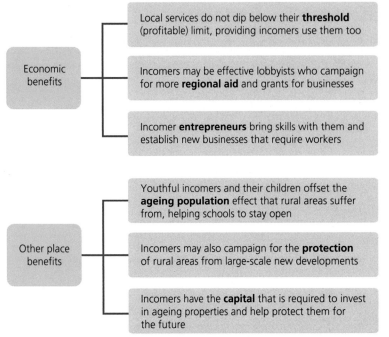

Economic benefits
- Local services do not dip below their **threshold** (profitable) limit, providing incomers use them too
- Incomers may be effective lobbyists who campaign for more **regional aid** and grants for businesses
- Incomer **entrepreneurs** bring skills with them and establish new businesses that require workers

Other place benefits
- Youthful incomers and their children offset the **ageing population** effect that rural areas suffer from, helping schools to stay open
- Incomers may also campaign for the **protection** of rural areas from large-scale new developments
- Incomers have the **capital** that is required to invest in ageing properties and help protect them for the future

▲ **Figure 6.12** In-migration can have many benefits for rural places providing there is community cohesion among incomers and established residents

counterurban and international migration, many of whom have struggled to find work since agriculture ceased to be a major employer. One pattern that has emerged is the high proportion of people in these communities who have supported the UKIP political party and voted in 2016 for the UK to leave the European Union. Research has shown that many rural-born people in Lincolnshire and East Anglia felt 'left behind' by the UK Government.

There are, of course, clear upsides to rural in-migration, as Figure 6.12 shows. An increase in population numbers may be essential for the survival of some public services, including schools, doctors' surgeries and post offices. If population numbers fall below a threshold critical limit, such services can be lost. Accelerated and potentially irreversible social and economic decline may follow. This mirrors the changes experienced in some urban areas on account of deindustrialisation (see page 85). In such cases, population growth is very much a good thing in certain respects.

ANALYSIS AND INTERPRETATION

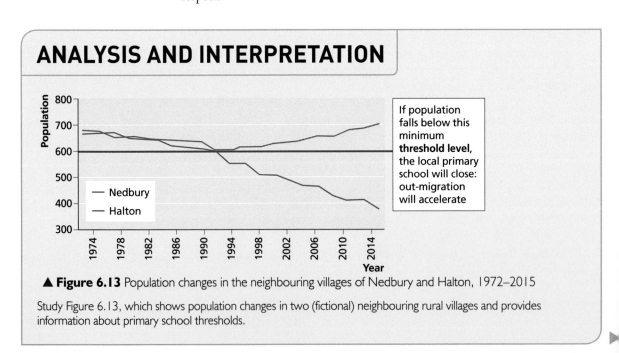

▲ **Figure 6.13** Population changes in the neighbouring villages of Nedbury and Halton, 1972–2015

Study Figure 6.13, which shows population changes in two (fictional) neighbouring rural villages and provides information about primary school thresholds.

(a) Estimate the difference in population size for Nedbury and Halton in (i) 1994 and (ii) 2010.

(b) Suggest why population changes for the two villages are similar at first but begin to differ over time.

GUIDANCE

This question is testing understanding of the threshold concept. Nedbury's population never declined below a critical threshold level that would have required the school to close. As a result, the village was probably able to attract new in-migrants from the 1990s onwards, including young families escaping city life. By 2014, village population had recovered to the same level as the early 1970s. In contrast, population loss in Halton was marginally worse than in Nedbury during the early 1990s. But this slight difference affected Halton catastrophically: its primary school would have closed when village population fell below the threshold level required for an educational service to be provided. In the years since then, even more families have left Halton because it has no school. Potential in-migrants avoid this place for the same reason. Halton has entered an unsustainable state whereas Nedbury has not.

(c) Explain possible actions which could be taken to help a declining settlement such as Halton.

GUIDANCE

This question provides an opportunity to demonstrate applied knowledge and understanding of the place-remaking process and its various strands, including regeneration, reimaging and rebranding. Given the small size of the settlement – fewer than 400 people – it is important to provide a realistic answer. This is not a settlement where a large flagship development would ever be built. Any permanent catalyst for redevelopment will need to be small-scale, given the context, such as a new farm shop. Alternatively, Halton could perhaps be chosen as the site for a new annual music festival or selected as the setting for a television show or film.

③ Place remaking in a rural context

▶ *How far have attempts at rural place remaking succeeded?*

Rural rebranding and regeneration

Table 6.1 shows a variety of place-remaking approaches and diversification strategies for rural areas. The most ambitious of these aim to raise the profile of places within a highly competitive globalised tourist industry; others seek local or regional success only. Many rural places are now significant tourist hubs thanks to intelligent management of their physical and cultural resources for tourism. However, there is also an argument for protecting some rural landscapes from excessively commercialised tourist flows and treating them instead as wilderness areas. This theme is returned to later in the chapter.

Strategy	Examples
Local heritage, including local history and monuments	■ Rural Britain is replete with local heritage attractions and events, from the annual Up Helly Aa fire festival in Shetland to Arthur's Quest at Land's End. ■ A facsimile Victorian village has been created at Shropshire's Ironbridge Gorge, a World Heritage Site widely recognised as the birthplace of Britain's Industrial Revolution. ■ Since 2017, rural tourism in ex-mining regions has benefited from a series of tours by Man Engine, a colossal mechanical tin miner (see Figure 6.14). This is a fine example of an updated visitor experience for heritage tourism (see page 131).
Natural heritage	■ A 96-mile stretch of southwest England was rebranded as The Jurassic Coast in 2001 after winning UNESCO World Heritage Site status for its geology and fossils. Subsequently, local businesses have leveraged funding for new and improved visitor attractions from a range of government sources and the National Lottery. ■ Another updated visitor experience is the glass viewing platform and dining area overlooking How Stean Gorge at Lofthouse in the Yorkshire Dales; it is available for party hire.
Traditional industries, crafts and food production	■ The market town of Ludlow, Shropshire, has built a reputation around food to maximise its potential for tourism. A network of local businesses has worked with town, county and district councils to achieve this goal. ■ The British countryside is home to many small-scale post-Fordist industries (see page 135), including textiles, wood and metal crafts, food and drink producers.
Farm diversification	■ The Edwards family of Ormskirk, Lancashire have rebranded their farm as Farmer Ted's Adventure Farm. Pre-school play areas and a petting zoo can now be found alongside traditional dairy farming. The change has been achieved with assistance from the local tourist board and education authority (see Figure 6.15). ■ The same farm has diversified further by staging a nightly 'Farmageddon' event during the Halloween season. Local people pay up to £35 (2017 prices) to be chased through the farm in the dark by zombies: visit www.farmaggedon.co.uk.
Media places	■ See page 142 for an account of media places, many of which are rural. Figure 4.23 (page 142) shows visitors in a rural media landscape setting (where TV show *Game of Thrones* was filmed).

▲ **Table 6.1** Rural rebranding and regeneration strategies

▶ **Figure 6.14** Man Engine, a colossal mechanical tin miner built with National Lottery funding, has toured rural mining regions of the UK, including the Cornish Mining World Heritage Site and South Wales (shown here visiting Swansea in 2018)

Pearson Edexcel

AQA

OCR

WJEC/Eduqas

Individual enterprises in rural places are often relatively small in size and, by themselves, would not have a large sphere of influence. However, some areas of the British countryside have been collectively reimaged by clusters of small enterprises to create post-productive tourist destinations with a large sphere of influence. An example is the county of Monmouthshire, which:

- won the 2008 Top Food Destination in Wales award on account of the large number of artisan food producers located in and around Monmouth; television presenter Kate Humble has diversified a farm ('Humble by Nature') near Monmouth where cookery courses are now available
- has long been popular with tourists due to its beautiful countryside and sites of historic interest (there are nine castles to visit near the town of Monmouth itself)
- contains the market town of Abergavenny, which now hosts an annual food festival and is close to the Green Man music festival, both of which attract visitors from all over the UK.

As a result of so many different rural enterprises – collectively catering to a wide range of diverse consumption interests – Monmouthshire can be viewed as a successfully reimaged rural area.

Not every rural venture succeeds though. The Canolfan Cywain Centre in Bala, North Wales, was 'flawed from the outset', an investigation by the Wales Audit Office found. The centre – which was supposed to showcase local heritage, create jobs and serve as a community arts centre – closed three years after it opened at a loss of £3.4 million for local enterprise agency Antur Penllyn. Evidence showed this redevelopment was likely to struggle from the start because of:

- flawed income assumptions (estimates of 40,000 visitors a year turned out to be wildly optimistic given the centre's remote location in central Wales)
- lack of clarity over what the centre was meant to offer, both to local people and visitors.

These failings meant an inability to attract private-sector funding and spark local multiplier effects.

▲ **Figure 6.15** Farmer Ted's Adventure Farm is an example of rural diversification; the business is reliant on flows of visitors and school trips from a relatively small local catchment area

CONTEMPORARY CASE STUDY: REBRANDING ARRAN

This case study explores issues surrounding the regeneration, reimaging and rebranding of the rural island of Arran in Scotland. It provides a valuable plenary illustration of changing places for three reasons.

1 New advertisements for Arran differ greatly from those used in the past, but these changing place representations have become a cause of tension and conflict for the island community.

2 The formal VisitArran rebranding plan that was launched in 2007 established visitor and financial targets, which means its success can be judged with reference to official benchmarks.

3 Several external shocks occurred in the decade after VisitArran was launched, including the global financial crisis and the Brexit vote; with this case study, we can explore the resilience of a rebranded tourist destination during tough times.

Changing times, changing tourism

Arran was traditionally marketed as a wilderness experience well suited for walking holidays. Typically, family groups holidayed for weeks at a time in cheap self-catering accommodation. The local tourist board promoted Arran using the slogan 'Scotland in miniature', which reflected the island's diverse geology, soils and vegetation. Arriving by ferry, visitors would typically climb the mountain of Goatfell, watch for wild deer or explore local geological features like Hutton's Unconformity. A 1965 *National Geographic* magazine article about Arran captured the way many people – locals and tourists alike – felt about the island: 'No one ever tried to turn her into a conventional resort. The native islander's attitude was "take it or leave it", an approach that preserved the island's beauty unsullied the way we visitors wanted it.' 'Unsullied' is an important word: it implies that commercial development spoils a place.

By the 1990s, the 'Scotland in miniature' slogan had run out of steam and action was needed to raise visitor numbers. Under the old paradigm, self-catering hill-walkers injected relatively little money into the local economy. To make matters worse, cheap budget flights to foreign destinations meant fewer UK citizens were now interested in domestic holidays. Full-time year-round employment had begun to vanish from the

island, threatening community sustainability. The case for providing wet-weather visitor facilities – where visitors would spend money all year round – had become compelling, along with the argument that Arran needed to change its image.

In the early 2000s, several large firms formed a business network with the shared goal of 'rebooting' tourism. Key players included the whisky distillery at Lochranza and the Auchrannie hotel spa resort. Together, they formed a Destination Management Organisation (DMO), which successfully gained funding from government agency Scottish Tourism. The DMO announced its objective to 'bring together businesses with public-sector organisations to collectively promote and manage the tourist industry on Arran'. There are clear parallels here with the successful work done by the Mersey Basin Campaign (see Chapters 4 and 5) in bringing together players in northwest England. Specific DMO actions included:

■ setting a benchmark target goal of 5000 extra visitors per year (see Figure 6.16)

■ developing a strong brand identity by promoting Arran as a short-stay premium destination boasting luxury establishments at Auchrannie and Lochranza

■ launching a new website called VisitArran to promote the island as an ideal short-break holiday and shopping destination: 'As soon as you step foot on Arran you'll experience events, activities and produce unique to the island,' promises the website

■ employing a PR company to stage UK consumer public relations exercises including the launch event that featured Miss Scotland (Aisling Friel) in a Hawaiian-style skirt with the iconic peak of Goatfell in the background (this look was called the 'Arran Aloha' – a quirky hybrid mix of Scottish and tropical ingredients – and perfectly reinforced the new 'island time in no time' official slogan adopted by VisitArran).

Figure 6.17 compares the 'Arran Aloha' campaign photograph with a 1960s postcard of the island. Chapter 2 (see page 51) introduced the *mise-en-scène* technique of image analysis that looks critically at how objects have been strategically arranged and framed in an image. Arran's mountains occupy centre stage as 'the main event' in the 1960s image (mountaineers are shown too, though only in a supporting role) but are demoted to being a mere background feature in 2007. This is completely in line with the way Arran is now widely promoted as 'a place to shop and dine in a scenic setting'. A critical analysis of the Arran Aloha might additionally consider the extent to which this is a place representation targeted at male heterosexual tourists. What's your view?

Evaluating the level of success

The 2007 launch event was a great success. Forty newspapers and 22 television shows reported on the rebranding of Arran, including the BBC, *The Sun*, *The Daily Mirror* and *The Daily Express*. In the years immediately afterwards, benchmark goals were met comfortably.

- Tourist spending reportedly grew from £27 million in 2006 to £35 million in 2010. Visitor numbers increased by 5000 and rose further again in 2015 when ferry company Caledonian MacBrayne reduced fare prices to the island.

- In 2015, 2016 and 2017, VisitArran and Arran tourist businesses won a string of Scottish tourism awards.

- Visitor numbers did fall briefly in 2009 following the global financial crisis. Arran was more resilient than many other destinations, however. Some experts are currently predicting a dire future for Scottish tourism as a whole because of possible shortages of eastern European workers. Perhaps Arran will demonstrate resilience again should these fears prove grounded.

Not everyone in Arran is happy with the changing direction of island tourism, however. Some older Arran residents preferred life when there was a quiet season on the island. This is particularly true for many English counterurban migrants who relocated to Arran precisely because of the unspoilt sense of wilderness which *National Geographic* once applauded. These people think the island would have been better served seeking protected National Park status instead of greater commerce. Local online newspaper, the *Arran Banner*, provides qualitative evidence of conflict. One letter-writer complained that 'a rather expensive hotel is not the *real* Arran experience'. Another regretted that 'Arran life used to be fantastic when roads were empty and there were less blinking tourists holding up traffic by driving at two miles an hour'.

Welcome to the *Isle of Arran*

▲ **Figure 6.16** 'Pay to play' leisure activities such as those on offer at the Balmichael Centre have helped Arran reach its increased visitor target

▲ **Figure 6.17** A 1960s postcard advertising Arran and the PR photograph that launched the 2007 VisitArran reimaging campaign. There is a marked difference in how the island's scenery is used

④ Evaluating the issue

▶ *Discussing differing views about the identity of rural places.*

Identifying possible views and rural contexts

In this final debate, we return full circle to the important themes of place meanings and representations explored in Chapters 1 and 2. Humans perceive, engage with and form attachment to places in ways that are bound up with their own different perspectives and experiences. The post-productivist economic 'vacuum' analysed earlier in this chapter has given rise to a 'polysemic' rural landscape. This means it has varied meanings and identities for different groups in modern society (see Figure 6.18).

- A first perspective is that some rural places should be set aside from human use completely and left to develop a wilderness identity. A case can also be made that rural areas which have been modified by centuries of human activity should now be 'rewilded'. This preservationist view is incompatible with development processes of any kind.
- In contrast, newcomers to the countryside may have been drawn there by 'rural idyll' representations in popular culture, such as those shown in Chapter 2. People who move somewhere because they value its natural landscape and traditional way of life are more

likely to object to new development plans that threaten this identity. They may exhibit NIMBY ('not in my back yard') conservationist attitudes and try to block change using planning laws.

- For some people who have lived in rural places their whole lives, the land may still be seen primarily as an asset that helps them earn a living. By viewing the countryside's main identity as a working landscape, the case can easily be made that new homes and employment opportunities must take priority over environmental issues.

Referring back to the concept of the differentiated countryside (see page 191), the viewpoints shown here become easier or more difficult to support depending on what kind of rural place we are discussing. Edgelands in the rural–urban fringe cannot realistically be restored to a pristine wild state (whatever that might look like). However, a far stronger case can be made for attempting to rewild areas of 'paternalistic' or 'clientilistic' countryside where the economic and social sustainability of sparse communities is in any case far from secure.

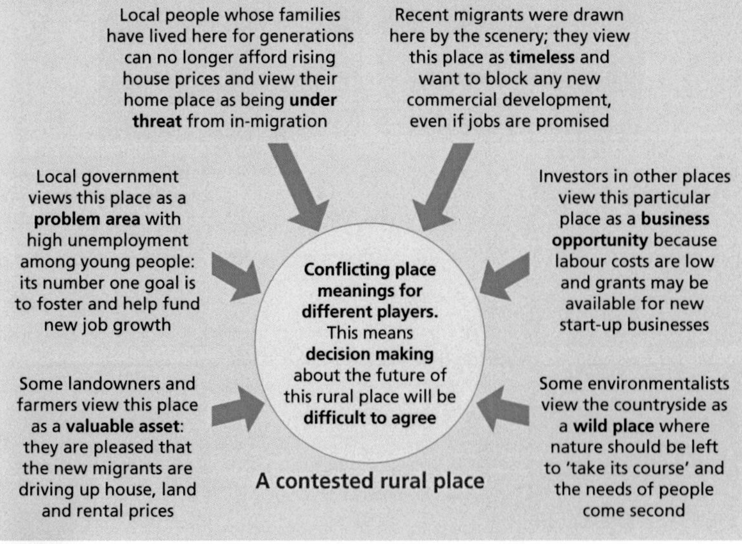

Local people whose families have lived here for generations can no longer afford rising house prices and view their home place as being **under threat** from in-migration

Recent migrants were drawn here by the scenery; they view this place as **timeless** and want to block any new commercial development, even if jobs are promised

Local government views this place as a **problem area** with high unemployment among young people: its number one goal is to foster and help fund new job growth

Investors in other places view this particular place as a **business opportunity** because labour costs are low and grants may be available for new start-up businesses

Conflicting place meanings for different players. This means **decision making** about the future of this rural place will be **difficult to agree**

Some landowners and farmers view this place as a **valuable asset**: they are pleased that the new migrants are driving up house, land and rental prices

Some environmentalists view the countryside as a **wild place** where nature should be left to 'take its course' and the needs of people come second

A contested rural place

◀ **Figure 6.18** Different groups of people have varying perspectives on rural identity (highlighted in bold) and the best way to manage rural places

View 1: some rural places are best left wild

A wilderness is a place with unspoilt natural characteristics (though views may differ on what constitutes 'natural'). Sometimes, a case will be made that a wilderness area under threat of exploitation should be set aside from development. In addition, there is an argument that some currently managed rural places should now be left alone and restored to their previous wilderness state. The word 'rewilding' is used to describe this process of allowing nature to take its course. A famous example of rewilding that you may have encountered previously was the reintroduction of wolves to the US Yellowstone National Park in 1995. Their return after 70 years of absence triggered an extraordinary trophic cascade (a linked sequence of ecosystem changes) and, more surprisingly, the restoration of landform systems including the re-establishment of trees and changes in the behaviour of river channels as their banks stabilised. All of this was triggered by the wolves reducing the number of grazing herbivores that had previously prevented the regrowth of trees.

The rewilding movement reflects shifting attitudes towards the natural world and how we view rural places and spaces. Many of us have come to expect environmental 'improvements' now that we belong to a post-industrial society, as the environmental Kuznets curve suggests (see page 97). Farming is less important as a source of wealth, while increased understanding of the ecosystem services concept has spread the view that the natural world has inherent economic value.

Restored peat moorland and forest ecosystems perform carbon sequestration services, for example. In the case of Yellowstone, the wolves delivered a natural service of population control by limiting deer and rabbit numbers. Rural places and organisms therefore gain a new identity as economic assets, even when they are left alone.

- Table 6.2 shows examples of rewilding in rural places throughout the UK. These schemes are mostly controversial. Farming unions have spoken out about farmers suffering financial loss from the reintroduction of beavers and the National Sheep Association is opposed to the reintroduction of lynx.

Place	Effect of rewilding
Norfolk and Cumbria (lynx)	1,300 years after Britain's last lynx was killed, the Lynx UK Trust wants to bring it back, arguing that big cats reduce deer numbers and restore ecosystem balance.
Argyll, Scotland (beavers)	Reintroduced beavers build river dams. These eventually drain to create meadows, thereby creating an important habitat for other species.
Isle of Mull, Scotland (white-tailed eagle)	In the 1980s, a breeding programme reintroduced the white-tailed eagle after 70 years of decline. Sixteen pairs are now established on the island.
Isle of Arran, Scotland (seabed restoration)	An example of marine rewilding, Scotland's first no-fishing zone was established in 2008. Lobsters, crabs, scallops and fish are all thriving in these 'wild' waters.

▲ **Table 6.2** Rewilding rural places in the UK

 KEY TERM

Ecosystem services Benefits for humans provided by ecosystems that would create economic costs if they became unavailable. Examples of ecosystem services include products such as food and water, regulation of floods and carbon storage.

- Ethical and philosophical dilemmas arise too. In Holland, a rewilded area called Oostvaardersplassen attracted criticism because horses were left to die when food became scarce in winter. Animal rights groups objected to 'nature taking its course' there. They argued that rewilding is really a form of scientific experiment. Logically, the horses are not truly wild and so, under law, deserve humane treatment.
- Finally, the concept of a 'natural identity' for rural places has been criticised, with some people asking: why is the end of the Pleistocene used as the 'base line' for rewilding? Further back, 115,000 years ago, the UK was home to elephants and rhinos. Why not make this the base line instead? Who has the power over these decisions and why?

View 2: rural places should be lived in but protected from identity changes

NIMBY feelings about rural identity often surface in relation to the attempted introduction of renewable energy sources in rural areas, particularly the installation of tall wind turbines. Some people view turbines as an asset for local areas, not to mention the country and world as a whole given our need to reduce humanity's carbon footprint. It is also possible to reimage an area using renewable energy; Chapter 4's rebranding ideas involving eco-places and technoscapes are applicable to rural places too. Northwest Cumbria – where many turbines are located – has been rebranded as 'Britain's energy coast' by a consortium of county councils and businesses.

There has been desperate opposition against turbines in other places because of their real and perceived impact on rural landscapes and place identity, however. In 2012, Bradford Council halted plans for development on the West Yorkshire moors near Haworth. This followed vigorous and ultimately successful campaigning by the Brontë Society.

- The proposed hundred-metre turbines, if allowed, would have been sited on the wild moors that inspired much of Emily Brontë's finest writing, including *Wuthering Heights*. The Brontë Society wished to conserve the identity of the moors exactly as it appears in the famous novels (see page 59).
- The chairperson of the Brontë Society said after the ruling: 'Visitors come from around the world come to see the wild moors of Emily Brontë's *Wuthering Heights* and want to see high waving heather – not high waving turbines. I am delighted by this decision and that all future applications will have to take into account the importance of the historical and literary associations of the area.'

An interesting issue that sometimes arises as part of the conservation process – and which was explored previously in Chapter 1 – is the vexing decision of where to actually establish a boundary for any rural area that needs conserving and protecting from identity change. One example is the decision to set a boundary for the South Downs National Park (SDNP), which finally became operational in 2011. Originally it was suggested that the SDNP should include a region called the Western Weald (see Figure 6.19). However, this place was excluded ultimately on the basis of differences in geology. The Western Weald lacks chalk parent rock, a defining characteristic of the SDNP. Many local people disagreed with that ruling, arguing that Western Weald nonetheless shared a general character with the rest of the rural region they perceive it as being embedded within. This argument was ultimately rejected.

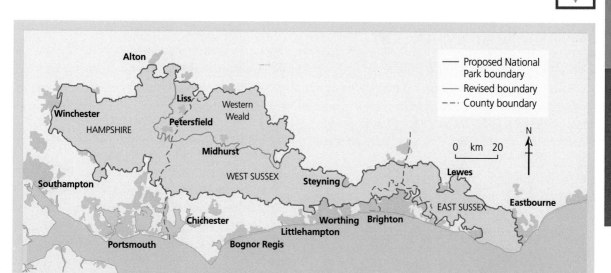

▲ Figure 6.19 The South Downs National Park boundary issue raised important questions about rural place identity: where does one place end and another begin?

View 3: new jobs and homes are needed so rural places must be allowed to change

A final view of rural place identity is that it should be allowed to evolve over time in accordance with the changing needs of local people. Wherever housing has become unaffordable due to demand among counterurban migrants, the case can surely be made that some land must be sacrificed to build affordable housing for those who need it, especially if we take the view that communities with long historical attachments to a particular place deserve a sustainable future *in situ*. The same case can be made for introducing new industries that will bring much-needed employment. Incomers to rural places often bring professional work with them on their laptops, but there may be limited job opportunities for long-established local families on account of the mechanisation of farming.

In 2017, a planning application was submitted by the company Treetop Treks for an 'Activity Hub' at Thirlmere in the Lake District. It included a proposal for eight zip wires across Thirlmere Lake (see Figure 6.20).

- Supporters of the proposals argued that the Activity Hub would create the equivalent of 28 full-time jobs, many of which would be year-round instead of just the summer season, and might attract over 100,000 visitors a year. In its planning submission, Treetop Treks said: 'The Lake District is not just for climbers and walkers. It is for everyone. It shouldn't be preserved in aspic.'

- The project faced very strong opposition from campaigners, who said the noise and sight of screaming riders would 'wreck the tranquillity' and spoil the identity of this area. A local group called 'Zip Off' collected thousands of signatures against the plan.

- There is a constant tension between development processes and conservation in this part of the Lake District, which is visited by 18 million people annually while also supporting 18,000 jobs.

Arriving at an evidenced conclusion

Why do views differ so greatly about the best way to manage rural identity? This discussion has explored a fundamental schism in the way different groups of people perceive, and engage with, rural places.

- Views are determined in part by aspects of someone's own personal identity, such as: their income bracket and employment prospects; whether they have migrated recently to a place or have deeper roots there; what their own personal and ethical beliefs are when it comes to protecting nature.
- Discussions and disagreements about rural identity intensify when there is no strong agreement about what the function of the countryside should be. In the past, most people subscribed to the norm that rural space should be used productively to generate resources, employment and wealth. Today, this view is still held by some people but other, often louder, voices have joined the debate.

- Inevitably, views about the importance of protecting rural identity vary according to what kind of rural place we are actually discussing. In previous chapters, certain landscapes were shown to have an iconic character because of the way they were once portrayed in literature, poetry, painting or film. Proposals to bring identity change to the world-famous Lake District will, of course, gather greater opposition than, say, a plan to modify Formby's edgelands, which far fewer people have heard of or care about.

In closing, it is important to remember that all rural identities will change eventually with the passage of time (regardless of what different people's views about change may be). The Lake District's glaciated troughs look very different now to the way they would have appeared 10,000 years ago shortly after the Pleistocene ice retreated. In years to come, climate change will most likely bring visible ecological and water-cycle changes to Thirlmere, irrespective of whether zip wires are introduced. No place's identity is secure in the long term. But this is not entirely a bad thing because it is change, along with continuity, that makes the study of places so very interesting for geographers.

◀ **Figure 6.20**
Thirlmere Lake: would screaming zip-wire riders change this place's identity irrevocably?

Chapter summary

✔ Rural places throughout the UK have been transformed by new flows of people, investment and resources. Some changes have been government-led (such as infrastructure projects), while others are people-led (such as counterurban migration).

✔ A wide range of players and organisations have helped reshape rural places through their collective governance over varying timescales, ranging from ephemeral festivals to more permanent changes in land use.

✔ Rural places can be classified in different ways. One approach is to distinguish between areas of preserved countryside, contested countryside, paternalistic countryside and clientelistic countryside. Rural places can also be categorised according to their economic and demographic characteristics.

✔ Demographic changes are an important influence on the survival of rural communities in the UK. Places with a shrinking and ageing population may cross a threshold of decline beyond which point important services can no longer be provided.

✔ Place remaking in a rural context includes rebranding, reimaging and regeneration strategies that often make use of the natural and cultural heritage found in different places. Scotland's Island of Arran is an example of successful rural rebranding, although there has been some local opposition.

✔ The management of rural places is often contested because perspectives differ on what the identity of the countryside should be. A spectrum of views ranges from support for rewilding of the countryside to calls for greater house-building in the future.

Refresher questions

1 What is meant by the following geographical terms? Wilderness; post-productive countryside; edgelands.

2 Using examples, outline the role played by different external forces in driving change in rural places.

3 Outline the characteristics of contrasting types of rural place you have studied.

4 Explain reasons why the importance of agriculture as a source of employment has declined over time in the British countryside.

5 Analyse recent economic, demographic and social changes in the population structure of some rural places.

6 What is meant by the following geographical terms? Threshold; positive feedback; sphere of influence.

7 Using examples, explain different kinds of rebranding strategy for rural places.

8 Outline arguments for and against the reintroduction of wolves and other animal species to rural places in the UK.

Discussion activities

1   In groups, discuss changes which the UK Government has forced upon rural places in recent decades. Should greater protection be given to rural places and communities?

2   In pairs, design a mind map showing possible synoptic linkages between the study of rural places and other geography topics such as: coastal and glaciated landscapes; the carbon and water cycles; tectonic landscapes and hazards; global systems and global governance.

3   In groups or pairs, carry out an assessment of the short-term and long-term impacts (both positive and negative) that festivals can have on rural economies and environments.

4   In groups, discuss the economic, demographic and environmental characteristics of rural places in the UK that you may have visited personally.

5   Discuss the view that all rural places in the UK have been affected by globalisation and global flows of people to some extent.

6   In groups, discuss the view that the housing and employment needs of rural people should always be more important than protecting the environment from development.

FIELDWORK FOCUS

The topics covered in this chapter provide many opportunities for an A-level independent investigation exploring rural places and issues.

A   *Creating an independent investigation exploring an aspect of the geography of music festivals.* You might be interested in analysing why a music festival such as Glastonbury, Green Man or Latitude is located in a particular place. A range of geographical factors such as land ownership, relief, drainage, soil type, flood risk, climate, transport networks and planning laws can be investigated. Alternatively, the impacts of a festival can be explored, ranging from the issues arising around recycling of site waste to a study of the festival's regional, national and global spheres of influence (interviews with festival-goers may reveal how far people have travelled). Alternatively, it would be interesting to find out how significant the benefits are for hotels and guesthouses in the local area.

B   *Investigating rural place characteristics and connections.* Most of the fieldwork suggestions in previous chapters can be adapted for rural place contexts. Table 6.3 sums up key ideas that could form the basis for a rural-based independent investigation.

Research focus	Possible questions
National influences on rural places	■ How important are transport policies for the survival of isolated rural places? ■ To what extent has broadband helped connect remote rural places with the rest of the UK?
Global influences on rural places	■ How globalised are rural places? ■ How do local rural people feel about the UK's decision to leave the EU?
Changing rural identities	■ How have rural people's incomes, health and education changed over time and why? ■ Has international migration resulted in a multicultural society in some rural places?
Representing rural places	■ How does the content of formal tourist websites compare with informal blogs and websites? ■ How different are rural place portrayals in texts created by local people compared with those written by tourists?

▲ **Table 6.3** Possible rural fieldwork themes

Further reading

ArranBanner.com. Available at: https://www.arranbanner.co.uk.

Cloke, P. and Little, J., eds. (1997) *Contested Countryside Cultures*. London: Routledge.

Cloke, P., Marsden T. and Mooney, P. eds. (2006) *Handbook of Rural Studies*. London: Sage.

Ford, R. and Goodwin, M. (2014) *Revolt on the Right: Explaining Support for the Radical Right in Britain*. London: Routledge: London.

Halfacree, K. and Boyle, P. (1998) Migration, rurality and the post-productivist countryside. In: H. Boyle and K. Halfacree, eds. *Migration into Rural Areas*. Chichester: John Wiley & Sons, 1–20.

Hatherley, O. (2014) Glastonbury: the pop-up city that plays home to 200,000 for the weekend. *The Guardian*.

Howells, H. (1965) Home to Arran, Scotland's magic isle. *National Geographic*, pp. 80–99.

Marsden, T., Murdoch, J., Lowe, P., Munton, R., and Flynn, A., eds. (1993) *Constructing the Countryside*. London: UCL Press.

Monbiot, G. (2013) *Feral*. London: Penguin.

Murdoch, J. and Marsden, T. (1994) *Reconstituting Rurality*. London: UCL Press.

Rae, A. (2017) *A Land Cover Atlas of the United Kingdom*. Sheffield: University of Sheffield.

Wood, M. (2004) *Rural Geography: Processes, Responses and Experiences of Rural Restructuring*. Sage: London

CHAPTER 7

Study guides

① AQA A-level Geography: Changing Places and Contemporary Urban Environments

Content guidance

The compulsory topic of Changing Places is supported fully by this book. Additionally, support for the optional topic Contemporary Urban Environments is provided by Chapters 4 and 5.

The Changing Places topic requires you to carry out an in-depth study of *two* places, one of which could be the home place where you live or study, or somewhere close by. The contrasting 'non-home' case study could be chosen from those featured in this book, for example Hampstead Heath, the Isle of Arran and the Isle of Dogs. It is important that the two case studies have contrasting economic or population characteristics. Study should be focused on:

- people's experience of these places and the meanings they have attached to them
- how these places have changed and developed over time
- relationships and connections between these and other places
- the range of qualitative and quantitative data sources that can be used to support study of these places, including statistical evidence and different media sources and place representations (for example, informal and formal sources).

The Contemporary Urban Environments topic requires study of urban processes, characteristics and management issues. This book supports the majority of the teaching and learning strands included in the AQA specification.

Changing Places (Topic 3.2.2)

This section of the AQA specification focuses on: 'How places are known and experienced, how their character is appreciated, the factors and processes which impact upon places and how they change and develop over time.' Detailed content is structured around four sub-themes.

Sub-theme and content	Using this book
3.2.2.1 The nature and importance of places The topic begins by establishing key concepts including insider and outsider perspectives of place, different categories of place (near places, far places, experienced places and media places). The character of places derives from a combination of endogenous factors (local physical factors, population characteristics and the built environment) and exogenous factors (relationships with other places).	Chapter 1, pages 1–18
3.2.2.2 Changing places – relationships, connections, meaning and representation Two contrasting case studies should be used to analyse important geographical understandings:	
■ Places are shaped by their relationships and connections with other people and places. These relationships impact on the demographic and cultural characteristics of places, and also economic and social changes and inequalities (you can choose to study *either* demographic/cultural changes *or* economic/social changes). You are expected to be able to analyse the importance of the different flows of people, resources and investment that help forge connections between places over time.	Chapter 1, pages 19–25 Chapter 6, pages 184–194
■ Places are impacted on by external forces operating at scales from local to global (you can choose to focus on the effects of *either* government policies, *or* TNCs, *or* the impact of global institutions).	Chapter 3, pages 74–97 Chapter 4, pages 117–149
■ Changes occur over time in the connections that link places together. Both past and present connections help to embed local places in larger-scale regional, national and global geographical contexts.	Chapter 1, pages 26–35
■ People attach important meanings and representations to places (both in the past and present). You should explore (i) how people perceive and form attachments to places in ways which reflect *their own* personal identities and perspectives, (ii) how external agencies attempt to create specific place meanings and thus shape the actions of other players and (iii) the diversity of different ways in which places are represented in various media.	Chapter 2, pages 39–58 and 66–69 Chapter 6, pages 195–199
3.2.2.3 Quantitative and qualitative skills A range of quantitative data, including geospatial sources, should be used to investigate and display place characteristics. In addition, you must be familiar with qualitative approaches and methods used to study and represent places. You are expected to be able to critically analyse the impacts of different representations on place meanings and perceptions.	Chapter 2, pages 59–65 Chapter 6, pages 198–199
3.2.2.4 Place studies It is important that all of the understandings listed above can be applied to two in-depth studies of (i) 'a place local to the home or study centre' (this could be your home or school neighbourhood) and (ii) 'a contrasting and distant' place study (either elsewhere in the UK or further afield). ■ As part of the study of your home place, it is important that you reflect on how your own life and experiences have been affected by 'continuity and change' in the place where you live. Do you, or other people you know, support a local sports team with a long history, for example? ■ Both place studies must focus on people's lived experience of the place in the past and present.	Possible case studies include: Hampstead Heath (page 18) Inner-city Sheffield (page 25) Hull (page 126) Salford Quays (pages 146 and 177) The Isle of Dogs (page 158) Central Birmingham (page 160) Formby (page 186) The Isle of Arran (page 198)

Contemporary Urban Environments (Topic 3.2.3)

This book additionally supports the study of contemporary urban environments, as shown in the table below.

Sub-theme and content	Using this book
3.2.2.1 Urbanisation	Chapter 1, pages 7–9
Topics include urbanisation, suburbanisation, counter-urbanisation, deindustrialisation and government regeneration policies for urban areas.	Chapter 5, pages 152–160
3.2.3.2 Urban forms	Chapter 3, pages 98–108
Topics include economic inequality, social segregation, cultural diversity and the factors that influence them. New urban landscapes should also be studied, including heritage quarters, fortress (gated) developments and gentrified areas.	Chapter 4, pages 131–136
	Chapter 5, pages 166–169
3.2.3.3 Social and economic issues associated with urbanisation	Chapter 3, pages 109–113
Issues to explore include economic inequality, social segregation and cultural diversity, along with strategies to tackle these issues.	Chapter 5, pages 161–165
3.2.3.5 Urban drainage	Chapter 4, pages 146–149
You should explore river catchment management and river restoration in an inner-urban catchment, with reference to a specific project (including the aims of the project, the involvement of different players and an evaluation of outcomes).	Chapter 5, pages 171 and 177–179
3.2.3.7 Contemporary urban environmental issues	Chapter 5, pages 170–175
Topics include atmospheric pollution, water pollution and urban dereliction, along with strategies to manage these environmental problems.	
3.2.3.8 Sustainable urban development	Chapter 5, pages 175–180
This section deals with the different environmental (natural/physical), social and economic dimensions of urban sustainability and how these are reflected in the features of sustainable cities. You should also explore the concept of liveability as part of an examination of contemporary opportunities, challenges and strategies for developing more sustainable cities.	

AQA assessment guidance

Changing Places and Contemporary Urban Environments are assessed as part of Paper 2 (7037/2). This examination is 2 hours and 30 minutes in duration and has a total mark allocation of 120.

There are 36 marks allocated for Changing Places. This consists of:

- a series of three short-answer questions (worth 16 marks in total)
- one 20-mark evaluative essay.

There are 48 marks allocated for Contemporary Urban Environments. This consists of:

- four multiple-choice questions worth 1 mark each (4 marks in total)
- one 6-mark question fully linked to a figure
- one 9-mark question partly linked to a figure
- one free-standing 9-mark question
- one 20-mark evaluative essay.

Changing Places short-answer questions (up to 16 marks)

Your first Changing Places question will most likely be a purely knowledge-based short-answer task targeted at assessment objective 1 (AO1) using the command word 'explain'. High marks will be awarded to students who can write concise, detailed answers which incorporate and link together a range of geographical ideas, concepts or theories. As a general rule, try to ensure that every point you make is either *developed* or *exemplified*:

- a developed point takes the explanation a step further (perhaps providing additional detail of how a process operates)
- an exemplified point refers to a relatively detailed or real-world example in order to support the explanation with evidence.

Your second short-answer question will make use of a figure and is targeted at assessment objective 3 (AO3). This means that you will be required to use geographical skills (AO3) to analyse or extract meaningful information or evidence from the figure. These questions will most likely use the command words 'analyse', 'compare' or 'assess'. The 'Analysis and interpretation' questions included throughout this book are intended to support the study skills you need to answer this kind of question successfully.

Your third and final short-answer question will again make use of a figure but is now targeted mainly at assessment objective 2 (AO2). It will use a command phrase such as: 'Analyse the figure and using your own knowledge ...' This means you are expected to use the data only as a 'springboard' to apply your own geographical ideas and information. For example, a 6-mark question – accompanying a text extract about the successful rebranding of a particular town – might ask: 'Using the figure and your own knowledge, assess the benefits rebranding can bring to places'. You can answer by writing about your own case studies and the benefits these places have gained.

Changing Places evaluative essay writing

The 20-mark Changing Places essay will most likely use a command word or phrase such as 'how far', 'assess the extent' or 'discuss'. The mark scheme will be weighted equally towards AO1 and AO2. For instance:

'Conflict often arises when people who live in a place try to resist changes that appear to have been forced upon them by organisations, groups and individuals from outside that place.' To what extent does this statement apply to one or more places that you have studied?

'Place and sense of place do not lend themselves to scientific analysis for they are inextricably bound up with all the hopes, frustrations and confusions of life.' With reference to both qualitative and quantitative data, assess the extent to which you agree with this quote.

Every chapter of this book contains a section called 'Evaluating the issue'. These have been designed specifically to support the development of the evaluative essay-writing skills you need to tackle tough questions such as these. As you read each 'Evaluating the issue' section, pay particular attention to the following.

- *Underlying assumptions and possible contexts should be identified at the outset.* For the questions shown above, before planning your answer think very carefully about what kinds of contrasting place contexts you could write about (urban or more rural places; small towns or large cities). 'Conflict' has a range of meanings (see page 66); so too does the phrase 'hopes, frustrations and confusions'. These ideas need to be carefully unpacked.

- *An essay needs to be carefully structured around different themes, views, scales, topic connections or arguments.* In the first essay, you could devise a structure where each paragraph deals with a different spatial scale ('outside' can have regional, national and global meanings). For essays which ask you to evaluate a viewpoint or quotation, such as the examples shown above, answers that score highly are likely to be well balanced in so far as roughly half of the main body of the essay will consider ideas and arguments supporting the statement; the remaining half deals with counter-arguments.
- *The command word 'evaluate' requires you to reach a final judgement.* Don't just sit on the fence. Draw on all the arguments and facts you have already presented in the main body of the essay, weigh up the entirety of your evidence and say whether – on balance – you agree or disagree with the question you were asked. To guide you, here are three simple rules.
  1 *Never sit on the fence completely.* The essay titles have been created purposely to generate a discussion that invites a final judgement following debate. Do not expect to receive a high mark if you end your essay with a phrase such as: 'So, all in all, conflict sometimes arises but at other times it does not.'
  2 *Equally, it is best to avoid extreme agreement or disagreement.* In particular, you should not begin your essay by dismissing one viewpoint entirely, for example by writing: 'In my view, conflict always arises when outside forces bring change to a place and this essay will explain all of the reasons why.' Instead, you should be considering different points of view.
  3 *An 'agree, but ... ' or 'disagree, but ... ' judgement is usually the best position to take.* This is a mature viewpoint which demonstrates you are able to take a stand on an issue while remaining mindful of other views and perspectives.

Contemporary Urban Environments questions

The principles set out above also apply to exam questions for the optional AQA topic of Contemporary Urban Environments. You will be required to answer one further 20-mark evaluative essay, such as:

'Deindustrialisation is the major cause of patterns of economic inequality and social segregation in urban areas in the 21st century.' To what extent do you agree with this statement?

Where the assessment of Contemporary Urban Environments differs notably from Changing Places is with the inclusion of two 9-mark questions.

- The first 9-mark question will be linked to a figure, is targeted at AO1 and AO2 and is likely to use a command phrase such as 'Using the figure and one other urban area you have studied ...'
- The second 9-mark question will be free-standing (it is not linked to a figure) and may include synoptic elements (see below).

Synoptic geography

In addition to the three main AOs, some of your marks are awarded for 'synopticity'. Instead of focusing on one isolated topic, you are expected to draw together information and ideas from across the specification in order to make connections between different 'domains' of knowledge, especially links between people and the environment (that is, connections across human geography and physical geography). The study of Formby's edgelands (see page 186) is a good example of synoptic geography; so too is the study of Salford Quays (see pages 146 and 177) because of the important linkages between water-cycle management and urban place remaking.

Throughout your course, take careful note of synoptic themes whenever they emerge in teaching, learning and reading. Examples of synoptic themes could include: the impact of tectonic hazards on urban architecture and population movements; strategies to reduce urban carbon footprint sizes in order to help mitigate anthropogenic emissions into the carbon cycle. Whenever you finish reading a chapter in this book, make a careful note of any synoptic themes that have emerged (they may have been identified or these could be linkages that you work out for yourself).

AQA's synoptic assessment

Some 9-mark or 20-mark exam questions may require you to link together knowledge and ideas from different topics you have learned about. These may appear in both your physical geography and human geography examination papers. For example:

- a water cycle question (paper 1) might ask you to think about ways in which water cycle changes could affect place characteristics
- a global systems question (paper 2) could ask you to discuss ways in which changing global trade flows have affected local place inequalities.

In the exam, your Changing Places essay might include a synoptic link. For example, look at the following question:

How far do you agree that future changes to local places will be driven primarily by changes in the Earth's global carbon budget rather than by any other processes of change?

The mark scheme would include the following statement: 'This question requires links to be made across the specification specifically between changing carbon budgets in the water and carbon cycles and the changing nature of places.' One way to tackle this kind of potentially tricky question is to draw a mind map when planning your response. Draw two equally sized circles and fill these with relevant ideas, processes and contexts, trying to achieve the best balance you can between the two linked topics.

Pearson Edexcel A-level Geography: Shaping Places (either Regenerating Places or Diverse Places)

Content guidance

Students following the Pearson Edexcel course are offered a choice of two topics under the general heading Shaping Places. Option 4A is called Regenerating Places and Option 4B is called Diverse Places. The material included in this book supports both (and there is in any case a considerable degree of overlap in content).

Both options require you to carry out an in-depth study of *two* locations, one of which is the home place where you live or study. Make sure you have done this. The second place could be a case study taken from this book, for example Hampstead Heath, the Isle of Arran or the Isle of Dogs. Contrasts should be drawn in terms of the way these two places:

- have characteristics which are influenced by the regional and national contexts they are embedded within
- are subject to global and international influences such as TNCs
- have experienced changes in identity over time, either economically or culturally
- are represented in diverse ways by formal and informal media
- are home to people with different identities whose varying attitudes towards local place changes may have led to tension or conflict.

In addition, both options require you to have consulted a range of data sources to assist with your place studies, including statistical evidence along with different media sources and place representations (for example, informal and formal sources).

Option 4A Regenerating Places

The focus of option 4A is the regeneration of places: this refers to state and private-sector attempts to attract new economic activity and investment to a local place or wider region. Throughout option 4A, students should reflect on how ideas about regeneration relate to the place where they live or study, along with a contrasting example of one other place. The content is structured around four main enquiry questions.

Enquiry question and content	Using this book
1 How and why do places vary?	Chapter 1, pages 1–35
The topic begins by exploring the four main sectors of economic activity (primary, secondary, tertiary and quaternary) and ways in which these vary spatially in their importance at varying scales within the UK, including the broad north–south divide that exists along with contrasts between neighbouring places within cities and rural areas. Additionally, changes in place functions and demographic characteristics should be understood. It is expected that you can explain place changes over time (on account of physical, historical or planning factors) and suggest ways of measuring these changes using census data, maps or other formal statistical sources. Finally, the links between place changes and past/present connections with other places, regions and countries need to be examined in some depth for the two case study places.	Chapter 3, pages 74–86
2 Why might regeneration be needed?	Chapter 3, pages 86–97
Contrasting successful and declining places and regions must be studied, with particular emphasis on (i) successful global hubs (such as San Francisco Bay, USA, or Cambridge, UK) that have attracted inward migration and investment, and (ii) declining 'rust belt' inner cities, such as Pittsburgh, USA, or Sheffield, UK, during the 1980s and 1990s. In both cases, you should understand how changes have affected people's perceptions of areas such as these and what the priorities for regeneration are in declining areas. In turn, this enquiry question requires you to study variations in the 'lived experience' of the different people (of varying age, ethnicity, gender, income, etc.) who inhabit these places and the ways in which their experiences are reflected in actions and levels of engagement with local issues. There are a range of ways of identifying and evaluating the need for regeneration in declining places.	Chapter 5, pages 152–156
	Chapter 3, pages 98–116
3 How is regeneration managed?	Chapter 4, pages 117–148
This begins with an in-depth look at national government decision making, including infrastructure investment, planning laws and national-scale decisions about migration and money markets (rules that govern foreign investment into the UK land economy). There is a focus too on local government decision making, for example ways in which local governments have stimulated the growth of science parks. You should be familiar with sports-led urban redevelopment, such as the London Olympics 2012, and rural diversification schemes. Examples of rebranding should be studied in both urban and rural contexts using a range of strategies, including heritage or outdoor pursuits. This part of the course also requires you to explore how regeneration requires different players to work together, which can sometimes lead to tension and conflict.	Chapter 5, pages 170–174
	Chapter 6, pages 195–197

Enquiry question and content	Using this book
4 How successful is regeneration? It is important you can carry out a genuine evaluation of success. A range of measures can be used, including economic data (providing an assessment of changing inequalities both *between* and *within* areas) and broader measures of 'liveability' (improvements in the living environment). Contrasting views on the changes brought by regeneration strategies form part of this enquiry question and it is important that you can look in some depth at the ways different groups have been affected by changes. Examples of both urban and rural contexts should be studied, along with contrasting stakeholder perspectives on regeneration or redevelopment.	Chapter 5, pages 156–169 and 175–180 Chapter 6, pages 198–204

Option 4B Diverse Places

The focus of option 4B is the diversity of different places. This refers to the level of heterogeneity (variation) exhibited by the communities of different places, measured in terms of demographic (population) structure, ethnicity, religion, language or other defining cultural traits. The content is structured around four main enquiry questions.

Enquiry question and content	Using this book
1 How do population structures vary? The topic begins by exploring variations in population structure between and within places at varying scales, including the UK's north–south divide and variations along the rural–urban continuum. Variations in fertility, mortality, migration and ethnicity should be recognised, analysed and explained (for example, with reference to reasons for migration and clustering, physical factors and the role of government policies). Finally, the links between place changes and past/present connections with other places, regions and countries needs examining in relation to the two selected case study places.	Chapter 1, pages 1–35 Chapter 3, pages 98–108 Chapter 6, pages 184–189
2 How do different people view diverse living spaces? There are a range of ways of identifying and explaining the varying experiences of places that different groups of people have, both historically and in the present. Important themes include (i) the way different urban places are perceived by residents or outsiders, and (ii) the diverse ways rural places are viewed by different groups (because of the way they are represented in the media, along with real physical challenges such as remoteness). The spectrum of urban and rural contexts should also be appreciated (for example, the contrast between inner city and suburbs, or the differences between remote rural areas and parts of the rural–urban fringe). There are a range of ways of identifying and evaluating differences in how people view their living spaces.	Chapter 2, pages 39–70 Chapter 6, pages 190–194
3 Why are there demographic and cultural tensions in diverse places? This begins with an in-depth look at increased cultural diversity in the UK and the spatial patterns that have developed (some places are far more culturally diverse than others). There is a strong focus here on levels of segregation and you need to be able to both describe and explain this phenomenon using a range of examples and indicators (including cultural landscape evidence such as ethnic retail outlets). This part of the course also requires you to explore how diversity issues can sometimes lead to tension and conflict among the many different players who create and manage cultural changes in urban environments.	Chapter 3, pages 74–85 and 98–113 Chapter 5, pages 161–163

▶

Enquiry question and content	Using this book
4 How successfully are cultural and demographic issues managed? A range of measures can be used, including the economic and social progress of different places and communities, along with cultural indicators such as the political engagement of different ethnic groups and actions to tackle racism and build community cohesion. It is important you can carry out a genuine evaluation of success. Crucially, social attitudes vary enormously in the UK about the success of policies addressing migration, multiculturalism and community cohesion (the UK's Brexit referendum was held after this course was written but this is clearly a key theme that students might benefit from being well-informed about). Finally, you need to explore diverging perspectives on the management of demographic and cultural change in rural places: not all people agree that rural areas should be managed in ways that encourage population growth and change.	Chapter 4, pages 120–134 and 143–148 Chapter 5, pages 161–169 and 175–180 Chapter 6, pages 198–204

Pearson Edexcel assessment guidance

Shaping Places is assessed as part of Paper 2 (9GE0/02). This examination is 2 hours and 15 minutes in duration and has a total mark allocation of 105. There are 35 marks allocated for whichever Shaping Places option has been studied (Regenerating Places is tested by question 3; Diverse Places is tested by question 4), indicating that you should spend around 45 minutes answering. The 35 marks consist of:

- a series of three or more short-answer questions (worth 15 marks in total)
- one 20-mark evaluative essay.

Short-answer questions

One or two of the short-answer questions will be targeted (in part) at assessment objective 2 (AO2).

- These questions will be linked to a figure and typically use the command word 'suggest'. For example, a 6-mark question – accompanying a map showing life expectancy variations within an urban area – might ask: 'Study the figure. Suggest why some places have a lower life expectancy than others'. To score full marks, you must (i) apply geographical knowledge and understanding to this new context you are being shown, and (ii) establish very clear connections between the question that is being asked and stimulus material you have been shown.
- The 'Analysis and interpretation' questions included throughout this book are intended to support the study skills you need to answer this kind of question successfully.

Note that the Pearson Edexcel examination of Shaping Places does *not* employ descriptive written assessment objective 3 (AO3) tasks such as 'describe the pattern shown in the figure' or 'analyse the trends shown in the figure'. However, you *could* be required to briefly complete a short skills-based numerical or graphical AO3 task. The specification includes a list of skills and techniques you are expected to be able to carry out, such as a Spearman's rank correlation test, the calculation of an interquartile range or accurate plotting of data on to a chart or graph.

You will also be asked one purely knowledge-based short-answer question targeted at assessment objective 1 (AO1). This will most likely be the first short-answer question using the command word 'explain'. High marks will be awarded to students who can write concise, detailed answers that incorporate and link together a range of geographical ideas, concepts or theories. As a general rule, try to ensure that every point you make is either *developed* or *exemplified*:

- a developed point takes the explanation a step further (perhaps providing additional detail of how a process operates)
- an exemplified point refers to a relatively detailed or real-world example in order to support the explanation with evidence.

Evaluative essay writing

The 20-mark essay will most likely use the command word 'evaluate' with a mark scheme that is weighted heavily towards AO2. For instance:

Evaluate the importance of rebranding to the success of rural regeneration. (Option 4A)

Evaluate the relative importance of local and national government decision-makers in the regeneration of either urban or rural areas. (Option 4A)

Evaluate the view that successful urban management for some is likely to be unsuccessful for others. (Option 4B)

Evaluate the contribution of both national and global influences to the cultural tensions in either urban or rural areas. (Option 4B)

Every chapter of this book contains a section called 'Evaluating the issue'. These have been designed specifically to support the development of the evaluative essay-writing skills you need to tackle tough questions such as these. As you read each 'Evaluating the issue' section, pay particular attention to the following.

- *Underlying assumptions and possible contexts should be identified at the outset.* For the questions shown above, before planning your answer think very carefully about what kinds of contrasting place contexts you could write about (fringe or remote rural places; small towns or large cities). What is meant by 'importance' and 'success' in these essay titles? Do the words 'some' and 'others' refer to individuals, groups of players or both? These ideas need to be carefully unpacked.
- *An essay needs to be carefully structured around different themes, views, scales, topic connections or arguments.* In the first essay, what else could be important other than rebranding? In the third essay, can you come up with a counter-argument showing that urban management has occasionally been carried out in ways which benefit everyone? For essays which ask you to evaluate a viewpoint, such as the examples shown above, answers that score highly are likely to be well-balanced in so far as roughly half of the main body of the essay will consider ideas and arguments supporting the statement; the remaining half deal with counter-arguments.
- *The command word 'evaluate' requires you to reach a final judgement.* Don't just sit on the fence. Draw on all the arguments and facts you have already presented in the main body of the essay, weigh up the entirety of your evidence and say whether – on balance – you agree or disagree with the question you were asked. To guide you, here are three simple rules.
  1 *Never sit on the fence completely.* The essay titles have been created purposely to generate a discussion that invites a final judgement following debate. Do not expect to receive a high mark if you end your essay with a phrase such as: 'So, all in all, the redevelopment of Liverpool city centre has been both successful and unsuccessful.'
  2 *Equally, it is best to avoid extreme agreement or disagreement.* In particular, you should not begin your essay by dismissing one viewpoint entirely, for example by writing: 'In my view, the governance of Salford Quays has been a total success and this essay will explain all of the reasons why.' Instead, you should be considering different points of view.
  3 *An 'agree, but ...' or 'disagree, but ...' judgement is usually the best position to take.* This is a mature viewpoint which demonstrates you are able to take a stand on an issue while remaining mindful of other views and perspectives.

Synoptic geography

In addition to the three main AOs, some of your marks are awarded for 'synopticity'. Instead of focusing on one isolated topic, you are expected to draw together information and ideas from across the specification in order to make connections between different 'domains' of knowledge, especially links between people and the environment (that is, connections across human geography and physical geography). The study of Formby's edgelands (see page 186) is a good example of synoptic geography; so too is the study of Salford Quays (see pages 146 and 177) because of the important linkages between water cycle management and urban place remaking.

Throughout your course, take careful note of synoptic themes whenever they emerge in teaching, learning and reading. Examples of synoptic themes could include: the impact of tectonic hazards on urban architecture and population movements; strategies to reduce urban carbon footprint sizes in order to help mitigate anthropogenic emissions into the carbon cycle. Whenever you finish reading a chapter in this book, make a careful note of any synoptic themes that have emerged (they may have been identified or these could be linkages that you work out for yourself).

Pearson Edexcel's synoptic assessment

Synoptic exam questions are worth plenty of marks and you need to be well-prepared for them. In the Pearson Edexcel course, an entire examination paper is devoted to synopticity: Paper 3 (2 hours, 15 minutes) is a synoptic 'decision-making' investigation. It consists of an extended series of data analysis, short-answer tasks and evaluative essays (based on a previously unseen resource booklet).

As part of your Paper 3 answers, you will need to apply a range of knowledge from different topics you have learned about and make good analytical use of the previously unseen resource booklet (the 'Analysis and interpretation' questions in this book have been carefully designed to help you in this respect). The context used in the resource booklet may well make use of themes drawn from Shaping Places, though for the sake of fairness it must be equally relevant to options 4A and 4B. For example, the diverging attitudes of different players about attempts to change a place (for example, by building a new airport, railway or reservoir) lend themselves particularly well to this kind of synoptic assessment. Specialist ideas and theories about place meanings could therefore form an important part of a Paper 3 answer.

③ OCR A-level Geography: Changing Spaces; Making Places

Content guidance

Students must study the compulsory topic of Changing Spaces; Making Places, which is supported fully by this book.

The Changing Places topic requires you to carry out an in-depth study of *two* places, one of which should ideally be the home place where you live or study. The contrasting 'non-home' case study could be chosen from those featured in this book, for example Hampstead Heath, the Isle of Arran or the Isle of Dogs. It is important that the two case studies have contrasting economic or population characteristics. Study of these and other places should focus on:

- people's attachment to, and connections with, places
- the dynamic multi-layered characteristics of places

- connections between places and globalisation, including shifting flows of people, money and resources
- patterns of inequality (the landscapes of 'haves' and 'have-nots')
- 'placemaking' projects (referred to in this book as *place-remaking* strategies) carried out informally or formally by individuals or institutions.

Changing Spaces; Making Places (Topic 2.1)

This section of the OCR specification explores 'the relationships and connections between people, the economy, and society and how these contribute to creating places'. The detailed content is structured around five sub-themes.

Enquiry question and content	Using this book
1 What's in a place? This introductory section examines the multi-faceted characteristics of places (including consideration of the human, physical and built characteristics) and the past and present connections and flows that helped shape place identity: these flows include people, resources, investment and ideas. You are expected to gain an understanding of the way local places are embedded in larger-scale regional, national and global geographies. Taken together, these ideas should be used to help create two contrasting place profile case studies (one of which is likely to be your home place).	Chapter 1, pages 1–31 Possible case studies include: Hampstead Heath (page 18) Hull (page 126) Salford Quays (pages 146 and 177) The Isle of Dogs (page 158) Central Birmingham (page 160) Formby (page 186) The Isle of Arran (page 198)
2 How do we understand place? Two key ideas feature in this section of the specification. First, people differ in their own identity (age, gender, ethnicity etc.) and this gives rise to different kinds of place experiences and attachments; globalisation and time–space convergence have also affected our sense of place (and space). Second, these feelings give rise to a variety of informal and formal place representations using many different kinds of media, ranging from graffiti to geospatial data.	Chapter 1, pages 32–35 Chapter 2, pages 39–70
3 How does economic change influence patterns of social inequality in places? Three key ideas feature in this section. First, social inequality – measured through indices such as education, employment and housing – varies between and within places. Second, global connections and globalisation have driven structural changes, deindustrialisation and the rise of the service industry, creating new opportunities as well as affecting patterns of inequality; government plays an important role in managing these changes. Third, you should study two contrasting case studies of places that can illustrate different types of evidence, causes and outcomes of social inequality.	Chapter 3, pages 74–97 and 109–113 Chapter 5, pages 161–169
4 Who are the players that influence economic change in places? You must be familiar with the range of different players capable of driving economic change, including governments, MNCs and international institutions. To support this understanding, a case study should be undertaken of one country or region that has been impacted by structural economic change. You should explore: the changing characteristics of the case study area; causes and impacts of changes; and the role of different players driving the changes.	Chapter 4, pages 143–148 Chapter 5, pages 152–160

Enquiry question and content	Using this book
5 How are places created through placemaking processes?	Chapter 4, pages 117–142
Three key ideas feature in the final section. First, placemaking can be carried out by a range of players of varying scales, including local community groups, planners, governments and other organisations. The overall aims may include attracting inward investment or broader community goals. Second, you should explore a range of place-remaking strategies, including reimaging, regeneration and rebranding (these could be in either rural or urban contexts). The role of different players must also be understood, including reasons why some groups contest (disagree with) efforts to rebrand places or change place meanings. Finally, a detailed case study is required of a place that has undergone rebranding. This study should include: the need for rebranding; strategies taken and the role of different players; the impact of rebranding on people's place perceptions; and its overall relative success.	Chapter 5, pages 170–180 Chapter 6, pages 195–204

OCR assessment guidance

Changing Spaces; Making Places is assessed as part of Paper 2 (H481/02). This examination is 1 hour and 30 minutes in duration and has a total mark allocation of 66. There are 33 marks allocated for Changing Spaces; Making Places, indicating that you should spend around 45 minutes answering. The 33 marks consist of:

- a series of three or more short-answer questions (worth 17 marks in total)
- one 16-mark evaluative essay.

Short-answer questions

Your first two short-answer questions will most likely be targeted jointly at assessment objective 3 (AO3) and assessment objective 2 (AO2). This means that you will be required to use geographical skills (AO3) to extract meaningful information or evidence from a resource, such as a photograph or chart; you will then be required to offer an explanation of this information using applied knowledge of places and place-remaking processes.

- These questions are likely to use the command word 'explain'. They will also include the instruction to 'use the figure' or 'study the figure'.
- For example, an 8-mark question (4 marks AO3, 4 marks AO2) – accompanying a map showing life-expectancy variations within an urban area – might ask: 'Study the figure. Using evidence, explain possible reasons why some places have a lower life expectancy than others'. To score full marks, you must (i) apply geographical knowledge and understanding to this new context you are being shown, and (ii) establish very clear connections between the question that is being asked and stimulus material you have been shown.
- The 'Analysis and interpretation' questions included throughout this book are intended to support the study skills you need to answer this kind of question successfully.

Your third short-answer question will most likely be a stand-alone knowledge-based task worth 6 marks targeted at assessment objective 1 (AO1) using the command word 'explain'. High marks will be awarded to students who can write concise, detailed answers which incorporate and link together a range of geographical ideas, concepts or theories. As a general rule, try to ensure that every point you make is either *developed* or *exemplified*:

- a developed point takes the explanation a step further (perhaps providing additional detail of how a process operates)
- an exemplified point refers to a relatively detailed or real-world example in order to support the explanation with evidence.

Evaluative essay writing

The 16-mark essay will most likely use a command word or phrase such as 'how far do you agree', 'assess' or 'discuss'. The mark scheme will be weighted equally towards AO1 and AO2. For instance:

'Placemaking is used by governments only to attract inward investment.' How far do you agree with this statement?

With reference to one or more places you have studied, assess the role a range of players can have in influencing economic change.

Every chapter of this book contains a section called 'Evaluating the issue'. These have been designed specifically to support the development of the evaluative essay-writing skills you need to tackle tough questions such as these. As you read each 'Evaluating the issue' section, pay particular attention to the following.

- *Underlying assumptions and possible contexts are identified at the outset.* For the questions shown above, before planning your answer think very carefully about what kinds of contrasting place contexts you could write about (fringe or remote rural places; small towns or large cities; local places or a wider region). 'Placemaking' and 'economic change' are big ideas that need careful unpacking. Governments can be local, national or international; players can be individuals, groups, governments or other organisations. Important parameters such as these should be established at the planning stage of your essay and may be mentioned in the introduction.
- *An essay needs to be carefully structured around different themes, views, scales, topic connections or arguments.* In the second essay, what different kinds of role are there and how and why might these roles vary in importance? In the first essay, can you come up with counter-arguments showing that placemaking may have alternative objectives? For essays which ask you to evaluate a viewpoint, such as the first example shown above, answers that score highest are likely to be well-balanced in so far as roughly half of the main body of the essay will consider ideas and arguments supporting the statement; the remaining half deals with counter-arguments.
- *Command words and phrases such as 'to what extent' and 'discuss' require you to reach a final judgement.* Don't just sit on the fence. Draw on all the arguments and facts you have already presented in the main body of the essay, weigh up the entirety of your evidence and say whether – on balance – you agree or disagree with the question you were asked. To guide you, here are three simple rules.
  1 *Never sit on the fence completely.* Essay titles are created purposely to generate a discussion that invites a final judgement following debate. Do not expect to receive a really high mark if you end an essay with a phrase such as: 'So, all in all, placemaking in Liverpool city centre has been both successful and unsuccessful.'
  2 *Equally, it is best to avoid extreme agreement or disagreement.* In particular, you should not begin your essay by dismissing one viewpoint entirely, for example by writing: 'In my view, placemaking is only really used by governments to attract investment and this essay will explain all of the reasons why.' Instead, you should be considering different points of view.
  3 *An 'agree, but ...' or 'disagree, but ...' judgement is usually the best position to take.* This is a mature viewpoint which demonstrates you are able to take a stand on an issue while remaining mindful of other views and perspectives.

Synoptic geography

In addition to the three main AOs, some of your marks are awarded for 'synopticity'. Instead of focusing on one isolated topic, you are expected to draw together information and ideas from across the specification in order to make connections between different 'domains' of knowledge, especially links between people and the environment (that is, connections across human geography and physical geography). The study of Formby's edgelands (see page 186) is a good example of synoptic geography; so too is the study of Salford Quays (see pages 146 and 177) because of the important linkages between water cycle management and urban placemaking.

Throughout your course, take careful note of synoptic themes whenever they emerge in teaching, learning and reading. Examples of synoptic themes could include: the impact of tectonic hazards on urban architecture and population movements; strategies to reduce urban carbon footprint sizes in order to help mitigate anthropogenic emissions into the carbon cycle. Whenever you finish reading a chapter in this book, make a careful note of any synoptic themes that have emerged (they may have been identified or these could be linkages that you work out for yourself).

OCR's synoptic assessment

In the OCR course, part of Paper 3 (Geographical Debates H481/03) is devoted to synopticity. For this exam, you will have studied two optional topics chosen from: Climate Change, Disease Dilemmas, Exploring Oceans, Future of Food and Hazardous Earth. In Section B of Paper 3, you must answer two synoptic essays worth 12 marks each (in total, this adds up to 24 marks).

Each synoptic essay links together the chosen option with a topic from the core of the A-level course, such as Changing Spaces; Making Places. Possible Paper 3 essay titles might therefore include:

Examine how the risks from tectonic hazards affect placemaking processes.

How far do you agree that climate change has become the greatest cause of change for places?

Examine how flows of people, money and resources can affect the transmission of communicable diseases between places.

One way to tackle these kinds of questions would be to draw a mind map to help plan your response. Draw two equally-sized circles and fill these with relevant ideas, processes and contexts, trying to achieve the best balance you can between the two linked topics. The mark scheme requires that your answer includes: 'clear and explicit attempts to make appropriate synoptic links between content from different parts of the course of study'.

WJEC and Eduqas A-level Geography: Changing Places

Content guidance

Both WJEC and Eduqas students must study the compulsory topic of Changing Places, which is supported fully by this book. You should learn about a range of places, including your home place. The case studies that feature throughout this book can all be used to support different parts of the Changing Places course. Study of these and other places should focus on:

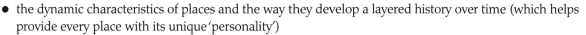

- the dynamic characteristics of places and the way they develop a layered history over time (which helps provide every place with its unique 'personality')
- the impact of economic restructuring and other external forces on local rural and urban places
- ways in which the home place and other case study areas are represented, for instance in tourist literature or the national media
- the way external forces at different scales, including businesses and governments, impact on the characteristics of places.

Changing Places

This section of both the WJEC and Eduqas specifications is structured around ten short sub-themes.

Enquiry question and content	Using this book
Changing places – relationships and connections This introductory section explores the demographic, socioeconomic and cultural characteristics of places; and the factors, flows and connections of people, resources, money and ideas that help shape these characteristics.	Chapter 1, pages 1–35
Changing places – meaning and representation This section deals with the meanings, perceptions and attachments that link people with places; and different ways in which places are represented in diverse media by formal and informal players. You should be aware of continuity and change in place meanings (that is the importance of heritage and history).	Chapter 2, pages 39–70
Changes over time in the economic characteristics of places Here, the focus is on economic structural changes driven by globalisation and the consequent changes in UK employment structure (as shown by the Clark–Fisher model).	Chapter 3, pages 74–78
Economic change and social inequalities in deindustrialised urban places The consequences of the loss of traditional industries include the cycle of deprivation as well as positive effects such as lower pollution levels. This section also explores government policies for deindustrialised places aimed at attracting investment and new forms of employment.	Chapter 3, pages 79–94 Chapter 5, pages 152–160
The service economy (tertiary work) and its social and economic impacts This section deals with the growth of retailing and commercial service-sector work driven by rising affluence and technological change. There are important links with gentrification in urban areas (where service-sector growth has attracted young professionals). Other service-sector issues include out-of-town retailing and internet shopping.	Chapter 3, pages 95–97
The 21st century knowledge economy (quaternary work) and its social and economic impacts You need to be able to describe and explain the growth of the UK's knowledge economy and the wider impact of this on places where knowledge (quaternary) industries have clustered (for example, university towns and cities).	Chapter 4, pages 136–141
The rebranding process and players in rural places This section deals with change and challenges in the post-productive countryside, including diversification and the subsequent reimaging and regeneration of rural places. You should also be familiar with the varying impact of these changes on different groups of people.	Chapter 6, pages 184–199

Enquiry question and content	Using this book
Rural management and the challenges of continuity and change Here, the focus shifts to rural places that continue to experience a range of challenges such as inadequate service provision, a lack of regeneration policies and housing shortages. You should also explore possible actions to manage these challenges.	Chapter 6, pages 200–204
The rebranding process and players in urban places This section explores place remaking in urban areas, encompassing reimaging, regeneration, flagship developments and the impact these strategies have on the actions and behaviours of different players.	Chapter 4, pages 117–135 and 143–148 Chapter 5, pages 175–180
Urban management and the challenges of continuity and change This plenary section reviews conflicts which may arise from the place-remaking process and the ongoing challenges for urban places where regeneration has not taken place or has caused 'overheating'.	Chapter 3, pages 109–113 Chapter 5, pages 166–169

WJEC and Eduqas assessment guidance

Changing Places is assessed as part of:

- *WJEC Unit 2*. This AS-level examination is 1 hour and 30 minutes in duration and has a total mark allocation of 64. There are 32 marks allocated for Changing Places, indicating that you should spend around 45 minutes answering. The 32 marks consist of:
  - two structured questions each worth 16 marks (32 marks in total), including two 8-mark 'mini-essays' that use a command word such as 'examine' or 'assess'.
- *Eduqas Component 1*. This A-level examination is 1 hour and 45 minutes in duration and has a total mark allocation of 82. There are 41 marks allocated for Changing Places, indicating that you should spend around 45 minutes answering. The 41 marks consist of:
  - a series of short-answer questions (worth 26 marks in total)
  - one 15-mark evaluative essay (from a choice of two).

Both courses use broadly similar assessment models and these are dealt with jointly below.

Short-answer questions

Some of the questions you will be asked will be linked to figures (maps, charts, tables or photographs).

- The opening question(s) will be targeted at assessment objective 3 (AO3). This means that you will be required to use geographical skills (AO3) to analyse or extract meaningful information or evidence from the figure. These questions will most likely use command words including 'describe', 'analyse' or 'compare'. The 'Analysis and interpretation questions' included throughout this book are intended to support the study skills you need to answer this kind of question successfully.
- Alternatively, you could be required to briefly complete a short skills-based numerical or graphical task. Your specification includes a list of skills and techniques you are expected to be able to carry out, such as a Spearman rank correlation test, the calculation of an interquartile range or accurate plotting of data on to a chart or graph.
- Finally, you may be asked to give a possible explanation of the information shown in the figure using applied knowledge. This kind of question is targeted at assessment objective 2 (AO2) and will most likely use the command word 'suggest'. It will also include the instruction: 'Use the figure'.
- For example, a series of short questions could accompany a map showing life expectancy variations within an urban area. The opening (AO3) question could be: 'Analyse the pattern shown in the figure'. The AO2 question which follows might ask: 'Suggest reasons why some places have a lower life expectancy than others shown in the figure'. To score full marks, you must (i) apply geographical

knowledge and understanding to this new context you are being shown, and (ii) establish very clear connections between the question that is being asked and stimulus material you have been shown.

Eduqas

In the Eduqas examination, some stand-alone short-answer questions are not accompanied by a figure. They are purely knowledge-based, targeted at assessment objective 1 (AO1) and worth up to 8 marks. They will most likely use the command words 'explain', 'describe' or 'outline'. For example: 'Outline reasons why it has become important to diversify the economies of rural places.' High marks will be awarded to students who can write concise, detailed answers which incorporate and link together a range of geographical ideas, concepts or theories. As a general rule, try to ensure that every point you make is either *developed* or *exemplified*:

- a developed point takes the explanation a step further (perhaps providing additional detail of how a process operates)
- an exemplified point refers to a relatively detailed or real-world example in order to support the explanation with evidence.

WJEC

In the WJEC examination, some short-answer questions are not accompanied by a figure. They are targeted jointly at AO1 and AO2 and are worth 8 marks. They will most likely use the command words 'examine' or 'assess'. For example: 'Examine reasons why it has become important to diversify the economies of rural places.' High marks will be awarded to students who can write concise, detailed answers which incorporate and link together a range of geographical ideas, concepts or theories. In addition, there should be some brief critical reflection, or weighing up of evidence, included as part of the answer. In the example above, an answer might briefly reflect on the different kinds of rural place this statement could be applied to (see page 191).

Evaluative essay writing (Eduqas only)

The 15-mark essays (10 marks AO1, 5 marks AO2) which feature in Eduqas Component 1 will most likely use the command words 'examine' and 'assess'. For instance:

Examine why some people benefit more than others from changes in the central areas of cities.

Assess how far different aspects of the rural rebranding process may rely on internet availability.

Every chapter of this book contains a section called 'Evaluating the issue'. These have been designed specifically to support the development of the evaluative essay-writing skills you need to tackle tough questions such as these. As you read each 'Evaluating the issue' section, pay particular attention to the following.

- *Underlying assumptions and possible contexts are identified at the outset.* For the questions shown above, before planning your answer think very carefully about what kinds of contrasting place contexts you could write about (retail districts or areas of housing in city centres; fringe or remote rural places).
- *Your essay can be carefully structured around different paragraphed themes, views or scales of analysis.* In the first essay above, think carefully about possible groups of people the essay could be structured around. Which different aspects of the rural rebranding process might you use to help structure an answer to the second essay? Important parameters such as these should be established at the planning stage of your essay and may be included in an introduction.

Synoptic geography

In addition to the three main AOs, some of your marks are awarded for 'synopticity'. Instead of focusing on one isolated topic, you are expected to draw together information and ideas from across the specification in order to make connections between different 'domains' of knowledge, especially links between people and the environment (that is, connections across human geography and physical geography). The study of Formby's edgelands (see page 186) is a good example of synoptic geography; so too is the study of Salford Quays (see pages 146 and 177) because of the important linkages between water-cycle management and urban place remaking.

Throughout your course, take careful note of synoptic themes whenever they emerge in teaching, learning and reading. Examples of synoptic themes could include: the impact of tectonic hazards on urban architecture and population movements; strategies to reduce urban carbon footprint sizes in order to help mitigate anthropogenic emissions into the carbon cycle. Whenever you finish reading a chapter in this book, make a careful note of any synoptic themes that have emerged (they may have been identified or these could be linkages that you work out for yourself).

The WJEC and Eduqas synoptic assessment

In the WJEC course, part of Unit 3 is devoted to synopticity while for Eduqas a similar assessment appears in Component 2. In both cases, synopticity is examined using an assessment called '21st Century Challenges'. This synoptic exercise consists of a linked series of four figures (maps, charts or photographs) with a choice of two accompanying essay questions. The WJEC question has a maximum mark allocation of 26; for Eduqas it is 30. Possible questions include:

Discuss the severity of the different risks that cities increasingly face.

To what extent could the management of different risks lead to changes in the characteristics of urban places?

As part of your answer, you will need to apply a range of knowledge from different topics you have learned about and make good analytical use of the previously unseen resources in order to gain AO3 credit (the 'Analysis and interpretation' questions in this book have been carefully designed to help you in this respect). One or both of the questions may relate quite clearly to the topic of Changing Places, as the example titles above demonstrate.

- Risks to cities could include the threats to employment and communities in urban places created by global shift and new technologies.
- Note also how the second question allows you to explore not just 'real world' architectural changes but also changing place *perceptions* (linked with feelings of increased danger or vulnerability).

Command words and phrases such as 'to what extent' and 'discuss' require you to reach a final judgement. Don't just sit on the fence. Draw on all the arguments and facts you have already presented in the main body of the essay, weigh up the entirety of your evidence and say whether – on balance – you agree or disagree with the question you were asked. To guide you, here are three simple rules.

1 *Never sit on the fence completely.* The essay titles have been created purposely to generate a discussion that invites a final judgement following debate. Do not expect to receive a really high mark if you end an essay with a phrase such as: 'So, all in all, there are many risks which cities face and they are all important.'

2 *Equally, it is best to avoid extreme agreement or disagreement.* In particular, you should not begin your essay by dismissing one viewpoint entirely, for example by writing: 'In my view, climate change is the greatest risk that all places face and this essay will explain all of the reasons why this is the case.' Instead, you should be considering different points of view.

3 *An 'agree, but … ' or 'disagree, but … ' judgement is usually the best position to take.* This is a mature viewpoint which demonstrates you are able to take a stand on an issue while remaining mindful of other views and perspectives.

Index

3D printing 95
actor/player networks 117, 146–147, 184–189
advertising 48–50, 53, 63, 120, 123–125, 132, 162, 198–199
affordable housing 34–35, 123, 128, 168–169, 172–175, 179, 185, 203
ageing populations 64, 90–91, 98, 101, 113–114, 179, 193–194, 205
agribusiness 190
air quality 170–171
Amazon 95–97, 118, 136
analysis iv
apprenticeships 85–86
AQA A-level study guide 208–213
Arran, Scotland 104, 198–199, 201
artificial intelligence 95–96, 178
artisan crafts 13, 19, 135–136, 197
assimilation 165–166
austerity measures 87–88, 110–111, 119, 163, 174, 179
Barcelona, Spain 30, 105
Barking and Dagenham, London 169
beavers 201
Belfast 68, 70, 142, 156–157
benchmarking sustainable development 176
BHS (British Home Stores) 96
bias 45, 51–52, 57–58, 65, 71–72, 109, 112
Birmingham 11, 19–20, 122, 134, 142–144, 160
Blache, Paul Vidal de la 6
Bloomberg headquarters 141
bombings 30, 34, 121, 125, 141
books & literature 12
    place change challenges 82, 93, 109
    place representation 48–49, 52–55, 58–59, 61–63, 67, 71–72
    rural places 186–187, 202
Bootle, Liverpool 6, 15, 112
bottom-up methods 33, 160, 166–167, 187
branding see rebranding
Brexit 42, 45, 82, 90–93, 95, 107–108, 159, 180, 198, 216
broadband 26–28, 63, 137–139, 160, 190–191
broken windows theory 84
Bruges, Belgium 33–34
brutalism 32–33, 35, 60–61, 122, 153
CAD see computer-aided design
CAM see computer-aided manufacturing
canals 22, 29, 32, 145–148, 170, 172, 177
Canary Wharf 60, 121, 128–129, 143, 145, 159, 173
capital 194
Capital of Culture, Liverpool 69, 124–126, 144
carbon neutrality 139–140
Cardiff 77, 108, 122–123, 128, 133–134

car manufacture 11–12, 29, 78, 80–82, 92, 135
Casey Review 165–166
central business districts (CBD) 111–112
Cereal Killer Café 169
Chelsea Harbour, London 167–168
Chester 20
China
    place change challenges 75–78, 80–82, 87–92, 97
    place dynamics/connections 19, 26, 29, 33
    place-remaking 127, 129, 144
    urban places 170–171
City of Culture 70, 124, 126, 138, 144–145, 151, 156
Clean Air Act of 170–171, 1956
Clerkenwell, London 6
clientelistic countryside 191–193, 200
climate 10, 43
climate change 9, 34–35, 93, 140, 204, 222, 226
clothing 43, 89, 96, 104, 135
cohesive communities 152, 161–169, 172–175, 181
Commonwealth Games 122, 125, 160
community cohesion 152, 161–169, 172–175, 181
community disintegration 167–168
commuting costs/times 172–174
computer-aided design (CAD) 80
computer-aided manufacturing (CAM) 80
conflicts
    place meanings/representation 39, 44, 48, 66–72
    rural places 184, 187, 191, 193, 198–200
    sustainable development 161–168, 181
connections/connectivity
    hyper-connectivity 57, 86–91
    place change challenges 106–107
    place connections 19–20, 24, 26–31, 36–37
    rural places 26–27, 29, 184–189
conservation processes 34, 139, 171–172, 200–203
construction, London 30–31, 47, 133
contemporary place-remaking 136–142
contested places 39, 47, 67–73, 129, 191–193, 200, 205
counterurbanisation 65, 84, 190, 193
counterurban migrants 185, 191, 193, 199, 203
Coventry 53, 69, 126
Crap Towns 12
creative class groups 166–169
Crossrail project 31, 47, 51, 133
culture

cultural diversity 14–15, 18, 21, 64, 74, 98–108, 113–114, 161–166
cultural heritage 131–136
cultural identities 41–45
cultural landscapes 10–14, 20, 28, 36, 47, 104, 131, 215
    place meanings/representation 39–66, 70–71
    popular culture 41–43, 47–49, 51, 53–55, 58–66, 68–69, 200
    sustainable development 161–169, 176
cumulative causation 152, 154–156, 159, 180–182
cycle of deprivation 74–86, 113–115, 118, 149, 164, 223
data manipulation skills iv
data sources 58, 109, 143
decanting 30, 133, 167–168
decarbonisation 91, 93–94
decoding meanings & messages 48–50, 71
defensive architecture 174–175
deindustrialisation
    place change challenges 74–86, 92, 97, 110, 113–115
    place dynamics 20–21
    place meanings/representation 53–54, 65
    place-remaking 118–119, 126, 135, 143, 149
    sustainable development 161, 163, 167, 170
demand thresholds 155
demographics see place demographics
demolition 19, 30–31, 47, 61, 120, 123, 145, 168, 179
deprivation, cycle of 74–86, 113–115, 118, 149, 164, 223
Devon 133, 138
diasporas 15, 37, 41, 128–129, 188
differentiated countryside 184, 190–195, 200
disagreements 39, 44, 48, 66–72
discourses 48, 50
discrimination 105, 164
diseases 22, 99, 101, 222
diversity, rural places 19, 184, 190–198, 214
Docklands, London 121, 133, 156–159
Dorset 99, 102, 133
drainage 171–172
dunes 186–187
Dyson Corporation 78
dystopian places 59–61, 65, 71–72
eagles 201
ecocities and eco-districts 137, 139–141, 149
ecological footprints 139, 178
E-commerce 19, 95–97, 112, 118, 136
economics
    economic functions 7–12, 14, 31, 36, 190

economic overheating 167–168, 180
growth 32, 35, 75, 86–89, 118, 123–126, 152–155, 161, 176
place change challenges 74–90, 94, 97–98, 102–103, 106–115
place-remaking 117–126, 131–139, 143–144, 147, 149
rural places 184–203
sustainable development 152–171, 175–183
ecosystems 43, 201
ecosystem services 201
Edexcel A-level study guide 213–218
edgelands 9, 186–187, 190, 200, 204
education 84–86, 99, 107, 109
Eduqas OCR A-level study guide 222–226
emerging economies 32, 35, 75, 88–89, 153
employment
place change challenges/inequalities 75, 77–89, 92, 95–96, 100, 102–103, 107–114
place characteristics 7, 11, 17
place dynamics 19, 22–27, 33
place-remaking 117–119, 123–138, 141–144, 146–149
place representation 53, 59–65
rural places 190–194, 197–198, 200, 203–204
sustainable development 153–159, 163, 172–180
encoding/decoding meanings/messages 48–50, 71
endogenous factors 19–22, 36, 188, 209
English Heritage see Historic England
enterprise zones 137, 156–160, 169, 181
entrepreneurialism 78, 92, 123, 126, 138–139, 154, 194
environment
Kuznets curve 97, 201
place and culture identity 44
rural places 184–206
stresses 170–180
sustainable development 152, 169–183
ethnic groups 104
ethnicity
place change challenges/inequalities 74, 103–106
place characteristics 15–16, 36–37
place meanings 50
place representation 54, 65–67, 70–71
sustainable development 161–165, 168
ethnoscapes 104
European Union (EU)
Brexit 42, 45, 82, 90–93, 95, 107–108, 159, 180, 198, 216
place-remaking 124, 127, 143–144, 148–149
evaluation iv
exogenous (external) factors 19, 21–23, 31, 36, 101, 119, 184–189, 209
experiences, place meanings/representation 39–48, 54–61, 70–72

export processing zones 75
Facebook 43, 50, 55–58, 72, 138, 145
fake place representations 57–58, 71–72
Farley, Paul 186–187
farming 13, 20, 23, 26, 140, 190–197, 200–203
far places 6, 19, 27–29, 36, 209
FDI see foreign direct investment
fearscapes 61–65
festivals 13, 161, 188–189, 197, 206
films, place-remaking 141–142
financescapes 60, 159
flagship developments 120–123, 133, 149, 153, 178, 224
food
place change challenges 101, 104
place-remaking 131, 140
place representation 43–44
rural places 190–191, 196–197, 201–202
football 11–12, 25, 29, 37, 82, 125, 128, 131, 143
Ford Motors 25, 81, 135–136, 149, 196
foreign direct investment (FDI) 26, 80, 92
foreign ownership, car manufacture 80–82
formal place representation 52–56, 63, 71
Formby, Liverpool 6, 15, 186–187, 204
foxhunting 67–68, 190–191
fracking 93
Friedmann, John 154
function, economic 7–12, 14, 31, 36, 190
gang warfare 68–69
gated neighbourhoods 167–168
GDP see gross domestic product
gender 62, 66, 70–71, 106
gentrification 15, 17, 120, 145, 166–169, 176, 179–182, 185, 193
geographic concept integration iv
geopolitical challenges 74–76, 83, 86–97, 106, 109–110, 113
GFC see global financial crisis
Ghost Town, The Specials, Coventry 53, 69, 126
Glastonbury Festival, Wiltshire 188–189
global agribusiness 190
global financial crisis (GFC) 76–78, 86–90, 110–114, 163, 174, 179–181, 198–199
global hubs 26, 76, 89, 112, 214
globalisation 206, 219
place change challenges/inequalities 113–114
place characteristics 5–6, 18
place connections 31, 35–36
place dynamics 19–23
place meanings 41
place representation 63, 71
global resource flows 21
global shift
place change challenges/inequalities 74–83, 86, 89, 97, 110, 113–114
place dynamics 19–20, 25, 36
place-remaking 121, 135–136
rural places 184
sustainable development 158

global stakeholders 128–130
Google Earth 57
governance 35
governments
economically sustainable development 152–161, 163–167, 171, 179–181
place-remaking 117–119, 126–131, 137–139, 143–150
rural places 185–188, 191–200
sustainable development 152–161, 163–167, 171, 179–181
Grade I, II* or II listed status 25, 33, 46, 61, 134
green belts 9, 34
green credentials 139–140
greenfield sites 33, 35
Green Man Festival, Brecon 189, 197
green politics 83, 112
Grenfell Tower, London 111–113, 168, 179
gross domestic product (GDP) 75–77, 87, 109
growth, economic 32, 35, 75, 86–89, 118, 123–126, 152–155, 161, 176
growth poles 153, 155–160, 167
Hallamshire 10, 25
Hampstead Heath 18
Hastings 133
healthcare 18, 22, 85, 98–99, 101, 155, 222
Hebden Bridge, Yorkshire 7, 69, 111, 193
heritage
place-remaking 120, 124–136, 141, 144–145, 148–151
rural places 188, 191–197
sustainable development 161, 164, 167
Hirschman, Albert 154
Historic England 33, 46, 61
history 164, 188, 191–197
Hogarth's cartoons 22
homelessness 174–175
home places 9, 31, 41, 53–57, 68, 114, 200
homogenous places 104–105
hotels 30, 104, 189, 193, 198–199
house prices 18, 58, 113, 116, 163, 167–168, 172–173, 179–180, 200
housing
affordable housing 34–35, 123, 128, 168–169, 172–175, 179, 185, 203
rural places 193, 200, 203–204
social housing 60–61, 111–113, 168, 179
sustainable development 152–159, 163, 166–182
Hull 70, 126, 138, 144–145, 148, 151, 156
hunting 33, 67–68, 71, 190–191
hyper-connectivity 57, 86–91
hyper-gentrified 167–169
identity
national identity 40, 45, 133, 165
places 5–18, 25–32, 39–47, 67–69, 81–83, 190–193, 200–204
image analysis 51, 198
India 75–76, 81–82, 87–92, 106–107

industries
  place change challenges/inequalities 74–116
  place characteristics 11–13
  place dynamics 19–29
  place-remaking 117–119, 123–138, 141–143, 146–149
  rural places 186–187, 189–198, 201–203
inequalities 74, 77–80, 108–117, 161–169
informal place representations 52–57, 69, 71–72, 120
inner-city challenges 83–86
Instagram 40–41, 55
institutional place representation 52–53, 55–56, 63, 71
interdependence iv-5
internal factors 19–22, 36, 188, 209
international migration 21, 102–106, 165, 184, 193–195
international stakeholders 128–130
Internet 14, 26–28, 40–44, 50–58, 69–72, 137–139, 145, 190–191
interpretation iv
investment
  foreign direct investment 26, 80, 92
  place change challenges/inequalities 79–83, 88, 92, 98, 111–114
  place remaking 120–129, 133, 138, 144–145
  place representation 65–67
  property asset purchasing 166–169, 179–180
  rural places 184–189, 194
  sovereign wealth funds 26, 88, 127–129, 144, 186
  sustainable places 152–160, 166–169, 172, 179–180
Isle of Arran, Scotland 104, 198–199, 201
Isle of Dogs, London 22, 156–160, 180–181, 213
John Lewis, Birmingham 144, 160
Keery, Alan and Gary 169
knowledge economy 137, 223
Kuznets curve 97, 201
Lake District 6, 55, 203–204
land reclamation 20
language 13, 21, 42–43, 79, 104–107, 165
later twentieth century deindustrialisation 74–86
LEP see local enterprise partnerships
life expectancy 15, 18, 21, 84, 97–103, 109–110, 216, 220, 224
listed status 25, 33, 46, 61, 134
literature see books & literature
liveability 152–153, 161–183
Liverpool
  Bootle 6, 15, 112
  Capital of Culture 69, 124–126, 144
  edgelands 186–187
  football club 29, 131
  Formby 6, 15, 186–187, 204
  industry 77, 84, 87, 112
  inequalities 112
  place-remaking 124–126, 130–132, 143–149

World Heritage Sites 34, 69–70, 124, 144
local area inequalities 111–113
local communities 145
local enterprise partnerships (LEP) 160
local places, importance in globalised world 41–42
London
  construction 30–31, 47, 133
  Grenfell Tower 111–113, 168, 179
  place characteristics/connections/ dynamics 6, 10–12, 15–24, 29–31, 34
  place-remaking 121–125, 128–133, 138, 140–145
  Soho 31, 47, 54
  sustainable development 153–154, 156–175, 179–181
low-carbon economies 91, 93–94
lynx 201
Malmö, Sweden 140
Malvern Hills 41
Manchester
  industry 77, 84, 99, 109, 112–113
  place-remaking 117, 121–126, 129–132, 139–149
  Ship Canal 145–146, 172, 177
Man Engine 196
manufacturing industries 11–13, 19–29, 74–114, 135–136
Masdar City, Abu Dhabi 140
MBC see Mersey Basin Campaign
meaning see place meanings
means of production 53
media
  media places 66, 69, 137, 141–142, 196, 209, 149.150
  place change challenges/inequalities 84, 109, 112
  place meanings/representation 41, 43, 47–72
  sustainable development 164, 169, 177–178
melting pot 105–106
Mersey Basin Campaign (MBC) 146, 177–178, 198
migration
  international migration 21, 102–106, 165, 184, 193–195
  place change challenges/inequalities 74, 76, 84, 94, 98–108, 111
  place characteristics/dynamics/ connections 6, 12–17, 21–22, 26–29, 32–34
  place identity 43–44
  place protection strategies 33–34
  place-remaking 129–130
  place representation 48, 64–66, 71
  rural places 6, 21–22, 32, 184–185, 193–195, 199–204
  sustainable development 155, 158, 163–167, 179
Milton Keynes 9, 136, 186
mining 82–83, 93–94, 113, 196
moral panic 61–62

multiculturalism 51, 105, 161, 164–165
multi-layered connections 28–29
multiplier effects 76, 92, 155, 159, 197
multi-scalar 127
Mumbai, India, place-remaking 140
museums 32, 46, 54, 60, 132–133, 143, 177–178
music 41–43, 47–49, 51–55, 59, 66–69, 188–189
music subcultures 54, 125
Myrdal, Gunnar 154
national grid 185
national identity 40, 45, 133, 165
nationalism 67–68, 70, 193
national resource flows 21
near places 6, 19, 27–29, 36, 209
negative externalities 170
negative feedback 156
neighbourhood inequalities 111–113
neoliberal 75, 119
Nestlé 78
Newcastle 88, 122–123, 148
Newham, London 162
New Street station, Birmingham 19–20, 160
NIMBY (not in my back yard) 18, 35, 179, 191, 200, 202
Nissan 80–82, 92
Northamptonshire 135–136
Northern Ireland 67–70, 128, 142
Northern Powerhouse 143, 148
north-south divide 77, 79–80, 99, 102, 108–113
nuclear power 93
nutrition 98, 101
OCR A-level study guide 218–222
offshoring 29, 75–76, 179
Olympic venues 30, 122, 153–154, 160, 214
online retailing 19, 95–97, 112, 118, 136
outsourcing 27, 75–76, 89
overheating issues 167–168, 180
palimpsest landscapes 32, 186
Park Hill estate, Sheffield 25, 32–33
partnerships in place-remaking 146–148
paternalistic countryside 191–192, 200
Peaky Blinders 62, 142, 153
perception changes, place connectivity 27–28
personal identities 40–45, 62, 71, 73, 161–162, 204, 209
physical elements, place-remaking 117, 120–158, 168–171, 176–178, 195–199, 205, 213–218
place attachments 39–42, 54, 72, 200
place change challenges/inequalities 74–116
place characteristics 5–19, 24, 31, 37
place connections 19–20, 24, 26–31, 36–37
place demographics 5, 209, 214–216
  place change challenges 74, 98–108, 111, 113
  place characteristics 11–19
  place dynamics 21, 24
  place protection strategies 33–34

rural places 184–187, 190–195, 205–206
  sustainable development 166, 179
place dynamics 19–25
place identity 5–18, 25–32, 39–47, 67–69, 81–83, 190–193, 200–204
place meanings 39–51, 64–72, 120–123, 131, 190–191, 200
place networks 26–31, 36–37
place protection strategies 32–35
place-remaking 117–158, 166–182, 193–205, 213–220, 224
place representations 39–40, 46–72, 109, 112, 129, 198–199
place retrofitting 137, 139
player networks 117, 146–147, 184–189
Plymouth 124, 133–134
politics
  place change challenges/inequalities 74–76, 83, 86–97, 106, 109–110, 113
  sustainable development 163–164
pollution 170–172, 177–178
popular culture 41–43, 47–49, 51, 53–55, 58–66, 68–69, 200
population see demographics
Port Talbot steelworks 88–91, 114
positive feedback 74, 83–86, 103, 113, 152, 155, 166–167, 169
post-accession migration 106, 108
post-colonial migration 21, 29, 106–107
post-Fordist activity 25, 135–136, 149
post-industrial 11, 16, 19, 97, 123–124, 143, 158, 166, 171, 201
post-productive 11, 19, 23, 185, 190–195, 197
poverty 62–63
preserved countryside 191, 198, 203
private-sector funding 144–145
property asset purchasing 166–169, 179–180
protection issues 32–35, 191, 194
psychological elements, place-remaking 120, 123–143, 146–151, 193–205
pull factors 64–65
push & pull factors 64–65
racial relationships 163–165
radicalisation 165–166
rationalisation 78, 90
rebranding 120, 123–125, 130–134, 137–138, 143, 148–151, 195–205
redevelopment 117, 120–158, 168–171, 176–178, 195–199, 205, 213–218
regeneration 117, 120–158, 168–171, 176–178, 195–199, 205, 213–218
regional aid 194
regional inequalities 77, 79–80, 99, 102, 108–113
regions 6–12, 15–17, 23, 25–27, 30–31, 36–37
reimaging 120, 123, 125–143, 146–150, 193–200, 202
religion 15, 42–44, 50, 67–70, 74, 104–106
re-making see place-remaking
renewable energy 93, 159, 202
reservoir construction 185–186

resilience
  place changes/inequalities 77, 80, 86–88, 95, 97, 110–113
  place protection strategies 35
resource flows 21, 184–189
retrofitting 137, 139
rewilding 200–202
rioting 145, 163–164, 169
rituals 13–14, 43
River Mersey 145–147, 172, 177–178, 198
river restoration/conservation 171–172
Roberts, Michael 186–187
robotics 95–96, 178
rough sleeping 174–175
rural diversification 19, 197, 214
rural gentrification 185, 193
rural idyll 62–64, 67–68, 71–72, 190–191, 200
rural places 184–207
  challenges 190–195
  characteristics 7–9, 11, 13–17
  connections 26–27, 29, 184–189
  differentiated countryside 184, 190–195, 200
  dynamics 19–23
  media representations 58–59, 62–68, 71–72
  migration 6, 21–22, 32, 184–185, 193–195, 199–204
  smart places 138–139
  sustainable development 167, 178, 191, 198, 200–203
rural–urban continuum 9, 190, 215
rural–urban fringe 9, 17, 186, 190, 192–193, 200
rural–urban migration 6, 21–22, 32
Salford Quays, Manchester 146–147, 172, 177–178, 180–181
sand shifts 186–187
Scarman Report 164
Scotland 44–45, 128–129, 138–139, 187–188, 191–193, 196–201
sea levels 18, 34
segregation 15–16, 104–105, 210
self-sustaining economic growth 152–154
sewage 146, 177
Shambles, York 28
Shankhill, Belfast 68–69
Sheffield 10, 21, 25, 32–33, 61, 84–85, 118, 131
shifting sands/dunes 186–187
shocks 75, 86, 97, 198
signifiers 40, 48, 63, 71
Silicon Roundabout 138
sites of places 10
smart cities 30, 89, 137–139
smart places 137–139, 143, 149, 155
smog 170–171
SDNP see South Downs National Park
snowballing 84, 103, 152–156, 159, 180–182
social housing 60–61, 111–113, 168, 179
social issues
  rural places 184, 190, 194, 200

sustainable development 156–158, 161–169, 172, 175–183
social media
  differentiated countryside 190–191
  place meanings 40–44
  place meanings/representation 40–44, 50–58, 66, 69–72
  place-remaking 137–138, 145
social networks 145
socioeconomics 14–15, 102–103, 182, 185, 223
Soho, London 31, 47, 54
soundscape 11, 54
sourcing data 58, 109, 143
South Downs National Park (SDNP) 202–203
Southport 186–187
sovereignty 67, 70
sovereign wealth funds (SWF) 26, 88, 127–129, 144, 186
space 5
spatial inequalities 74, 99–101, 109–113, 116
sphere of influence 41, 197
spread effects 156
Square Mile, London 141
stakeholders 117–119, 122–131, 137–150, 184–189
steel, Sheffield 25
steelworks/industries 10, 21, 25, 29, 76–78, 84–91, 114, 131
stresses, environmental 152, 170–175
structural economic changes, UK 76–86
study guides 208–226
subcultures 47, 54–55, 125
supercities 112
superdiversity 105
super-gentrified 167–169
sustainable development 152–183, 191, 198, 200–203
SWF see sovereign wealth funds
synoptic links iv
Tata Steel 89–90
technological changes
  place challenges 74–75, 80, 86, 89, 94–98, 113–114
  place connections 27–28
  place dynamics 22
  place-remaking 117–118, 131, 136–142, 149
technoscapes 47, 60, 141, 144, 202
television 142
tension
  place meanings/representation 39, 44, 48, 66–72
  rural places 184, 187, 193, 198, 200
  sustainable development 161–168, 181
terminal towns 88, 112
terror attacks 29
terrorism 29, 62, 165
texts, place representation 48–55, 59
three dimensional 3D printing 95
thresholds 55, 63, 85, 88, 93–97, 103, 155, 194–195, 205
tidal power 93
time–space convergence 22, 27–29, 219

tipping points 55, 63, 85, 88, 93–97, 103, 155, 194–195, 205
TNC *see* transnational corporations
tobacco industry 187
Tokyo, Japan 140
top-down methods 33, 160, 166–167, 187
tourism
    Barcelona, Spain 30
    place meanings/representation 48–55, 63, 69
    place-remaking 120, 123–124, 129, 131, 148–150
    rural places 187–199
    sustainable development 171–172
Toyota 81–82
trade unions 20–21, 82–83
transcripts 48
transnational corporations (TNC) 12, 26, 28, 43, 55–57, 75, 80–81, 92, 128, 136, 138, 144
transport 170–180
trickle-down 156–157
Trump, Donald 42, 119
turf warfare 68–69
Turning Basin at Salford Quays 146–147, 177
twenty-first-century
    challenges 86–97
    enterprise zones 159–160
Twitter 57, 145

UDC *see* Urban Development Corporations
unemployment *see* employment
United Nations Educational, Scientific and Cultural Organization (UNESCO) 32–34, 70, 127, 144
Urban Development Corporations (UDC) 156
urban entrepreneurialism 123
urban places
    characteristics 7–9, 11–12, 15–18
    connections 29–31
    construction 30–31
    deindustrialisation 74–86
    dynamics 19, 21–23
    farming 140
    place-remaking 117–151
    protection 32–35
    representation 58–62, 64–66, 68, 71–72
    rural–urban migration 6, 21–22, 32
    sustainable development 152–183
urban sprawl 9, 18
utopian places 30, 59, 71–72, 122, 137
Victoria Quays, Sheffield 32
violence 44–46, 61–62, 66–70, 83–84, 119, 132, 163–166
virtuous cycles 167
Wales
    Cardiff industry 77, 108

    place-remaking 122–123, 128, 133–134
    Port Talbot steelworks 88–91, 114
    rural places 193, 196–197
Walkie-Talkie building 141
warehouses 32, 118, 134, 187
wars 34
water quality 171–172
waterways 118, 145–148, 157, 172, 177
weather, extreme events 34
white elephants 153
WHS *see* World Heritage Sites
wicked problem 85–96, 168
Wikipedia 56–57
wilderness, rural places 184, 191, 195, 198–202
WJEC OCR A-level study guide 222–226
wolves 201
Worcestershire 136
World Heritage Sites (WHS) 32–34, 69–70, 124, 127, 133, 144, 196
Yorkshire
    Hebden Bridge 7, 69, 111, 193
    place characteristics 6–7, 22
    place networks and layered connections 26, 28
    place-remaking 136
YouTube 14, 43–44, 53, 55
zero-sum games 123–124, 157
zero tolerance measures 174–175

Photo Credits

p.6 © chrisdorney - stock.adobe.com; **p.7** © petejeff - stock.adobe.com; **p.9** © Justin Kase zsixz / Alamy Stock Photo; **p.12** *t* Sheffield United FC, *c* Arsenal Football club, *b* Derby county Football club; **p.14** *t* © OLI SCARFF/AFP/Getty Images, *b* ©Shutterstock / 1000 Words; **p.18** © I-Wei Huang - stock.adobe.com; **p.20** *t* © Nicholas bailey/REX/Shutterstock, *b* © Simon Oakes; **p.22** © Wellcome Library, London. Wellcome Images/http://creativecommons.org/licenses/by/4.0/; **p.25** © Simon Oakes; **p.28** ©antonel - stock.adobe.com; **p.30** © Simon Oakes; **p.32** © Simon Oakes; **p.33** © Simon Oakes; **p.40** © Archive Images / Alamy Stock Photo; **p.41** *t* © bloomua - Fotolia.com, *b* © David Martyn Hughes /123RF.com; **p.42** © Shutterstock / lesley rigg; **p.46** *t* © Ashley cooper / Alamy Stock Photo, *b* © Claudio Divizia - stock.adobe.com; **p.47** *l* © Ewan Munro/https://commons.wikimedia.org/wiki/File:Astoria,_Soho,_WC2_(2570670783).jpg/https://creativecommons.org/licenses/by-sa/2.0/deed.en, *r* © Jansos / Alamy Stock Photo; **p.48** © Simon Oakes; **p.49** *t* © Simon Oakes; *b* © StudioCanal/REX/Shutterstock; **p.50** © Donald McGill Archive; **p.54** © tupungato/123RF.com; **p.55** © pxl.store - stock.adobe.com; **p.60** *l* © Simon Oakes; *r* © Simon Oakes; **p.61** 2000 AD® is a registered trademark. Judge Dredd® is a registered trademark © 2018 rebellion 2000 AD ltd. All rights reserved. reproduced with permission of the copyright holder; **p.62** *l* ©classicpaintings / Alamy Stock Photo, *r* © Artisan Pics/Kobal/REX/Shutterstock; **p.68** *t* © Ellen Wallace/123RF.com, *b* © Jelle van der Wolf/123RF.com; **p.78** © Heiko Küverling/123RF.com; **p.82** © Roger Viollet/REX/Shutterstock; **p.83** © Trinity Mirror / Mirrorpix / Alamy Stock Photo; **p.84** © Matthew Ragen/123RF.com; **p.88** © MediaWorldImages / Alamy Stock Photo; **p.89** ©leighton collins - stock.adobe.com; **p.93** © Chris Dorney/123RF.com; **p.96** © Anthony Baggett/123RF.com; **p.105** © Simon Oakes; **p.112** © Simon Oakes; **p.115** Office for National Statistics. This information is licensed under the Open Government licence v3.0.; **p.118** © Ian Showell/Keystone/Getty Images; **p.121** © Simon Oakes; **p.122** bs0u10e0/https://commons.wikimedia.org/wiki/File:Selfridges_Building,_Birmingham_(2012).jpg/https://creativecommons.org/licenses/by-sa/2.0/; **p.125** © Shutterstock / kenny1; **p.126** © LEE BEEL / Alamy Stock Photo; **p.129** ©moramora - stock.adobe.com; **p.132** Steinsky/https://commons.wikimedia.org/wiki/File:070522_ukbris_ch01.jpg/https://creativecommons.org/licenses/by-sa/3.0/deed.en; **p.134** © Simon Oakes; **p.135** © Everett collection, Inc. / Alamy Stock Photo; **p.138** Tech City UK/Design by Smartup Visuals; **p.139** *t* © Roman Milert - stock.adobe.com, *b* © Simon Oakes; **p.140** © alisonhancock - stock.adobe.com; **p.141** *l* © Shutterstock / chrispictures, *r* ©zefart - stock.adobe.com; **p.142** www.fangirlquest.com; **p.145** © MEN Media; **p.148** © SakhanPhotography - stock.adobe.com; **p.153** © irstone/123RF.com; **p.154** © Mike Hughes/Alamy Stock Photo; **p.159** © Simon Oakes; **p.160** Birmingham City Council; **p.163** © Bob Dear/AP/REX/Shutterstock; **p.164** © Everett collection Inc / Alamy Stock Photo; **p.165** © Simon Oakes; **p.166** © Simon Oakes; **p.167** © Simon Oakes; **p.169** © alenakr/123RF.com; **p.171** *t* © Trinity Mirror / Mirrorpix / Alamy Stock Photo; *b* © Simon Oakes; **p.175** © Simon Oakes; **p.178** ©Debu55y - stock.adobe.com; **p.185** *t* © Giles/Daily Express/Express Syndication. Image from British Cartoon Archive, University of Kent, *b* © Sheffield Libraries and Archives / www.picturesheffield.com; **p.187** © CITiZAN; **p.188** © Anthony brown/123RF.com; **p.192** © Joe Dunckley / Alamy Stock Photo; **p.196** © Simon Oakes / with kind permission of Golden Tree Productions; **p.197** © Simon Oakes; **p.199** *l* & *c* © Simon Oakes, *r* © PA Archive/PA Images; **p.204** ©drewrawcliffe - stock.adobe.com